Recollections of
Garelochhead
100 Years Ago

by William Hamilton

Northern Bee Books

Recollections of Garelochhead 100 Years Ago
© 2017 William Hamilton

All rights reserved. No part of this publication may be reproduced, stored in a retrieval system, transmitted in any form or by any means electronic, mechanical, including photocopying, recording or otherwise without prior consent of the copyright holders.

ISBN 978-1-912271-09-2

Published by Northern Bee Books, 2017
Scout Bottom Farm
Mytholmroyd
Hebden Bridge
HX7 5JS (UK)

Design and artwork
D&P Design and Print
Worcestershire

Printed by Lightning Source, UK

Recollections of
Garelochhead
100 Years Ago

Foreword

My uncle, William Hamilton, was born in Garelochhead in 1889. He died in 1977 at his home at Blairmore. In his last few years he compiled this account of his younger days in Garelochhead which provides a fascinating insight into life in the village at the end of the nineteenth century and the beginning of the twentieth.

William, like his father, grandfather, and brothers, became a joiner after schooling. However, it was beekeeping which became his career. He acquired his first hive at the age of 14 and in his early thirties he was a lecturer in beekeeping at the West of Scotland Agricultural College. Some years later he was appointed lecturer in beekeeping at Leeds University and later a similar post at the Yorkshire Institute of Agriculture where he remained until his retirement in the early fifties. His book, 'The Art of Beekeeping' was first published in 1945 and became a best seller on the subject.

I have edited the original typescript and added a few footnotes. Considering that he was in his mid-eighties when he wrote these notes, his powers of recollection were remarkable. He has described his childhood here as the happiest days of his life. Certainly, he tends to dwell on happier events and rarely on the hardship that undoubtedly existed in the same period.

The photographs were not part of the original document but have been added for interest. Most of these were from old postcards collected by Alistair McIntyre, a well-known local historian who lives in Garelochhead. Alistair has been an enthusiastic supporter for the idea of publishing this book, and I am indebted to him for his help and advice throughout the process.

Graham J Hamilton

William Hamilton in Royal Flying Corps uniform.

Contents

Foreword

1. Early years; Queen Victoria's Diamond Jubilee etc.
2. The Blacksmith: Work and hobbies etc.
3. School dress, Sunday School, Barefoot boy etc.
4. Hogmanay, Sports, Cow, Hens etc.
5. The new road to Arrochar, Reliability trial etc.
6. Blanket washing; Sea Trout.
7. Natural history: (Birds and beasts on shore and woods).
8. Bendarroch; the Brownes; the Gibson Hall etc.
9. School; Mr. Connor; etc.
10. Food: Fish, meat, and milk.
11. Sanitation; Housing etc.; Dress of women and men.
12. South African War; Gaelic speakers; Tinkers; Irishmen; Turf dykes; Miss Grant of Grant.
13. The Roadman; First hive of bees etc.
14. Christmas trees; Donald the Shetland pony.
15. First grouse drive, and last one on Ardgoil.
16. Splashing for trout, and netting herring.
17. First game of bowls and how to be a Champion.
18. The Pier and the Steamers, the Surge, the West Highland Railway.
19. Miss MacDonald of Belmore; Faslane burial ground.
20. The Volunteers and Territorials, Rifle shooting and the first camp of 9th Argyll

Conclusion and Poem on Gare Loch

Early Years;
Queen Victoria's Diamond Jubilee etc.

One of the more pleasant aspects of growing old is to be able to look back to the days of one's youth and to relive the happy years of childhood and adolescence. As memory's pages are turned many of the early ones are indistinct or blurred or just blank leaves. But many are clear and distinct. Others, of little interest can be left undisturbed. It is fashionable to describe the era of my youth as the "bad old days" but to me they were and are the "good old days".

I was born in Garelochhead in 1889. I have the impression, indelibly imprinted in my mind, that the people, among whom I lived, worked, and played, for the first twenty years of my life, were contented with their lot. They had time for work and a time for play and entertainment, and they appeared to enjoy every waking moment. Doubtless they had their trials and tribulations but, looking back, I seemed to have lived in a dream world of happy and kindly folk. Perhaps I was fortunate in being born in such a place away from the squalor of towns, a place of natural beauty, unsurpassed in all Scotland, with a strong community spirit among the people, as will be seen in due course. Yet we were near enough to the big city, an hour journey away by train, to obtain the full benefits of modern progress.

I was born in Oakfield Cottages, in the smallest one in a row of three, consisting of two rooms. In the middle cottage, of four rooms, lived my old grandmother, three aunts and an uncle. I have only a hazy recollection of my early years but one or two incidents I do remember.

When I was about eighteen months old I had been taken for a day's sail on a steamer. I do not know who was with me. Suddenly I was awakened by someone, who kept shaking me. I opened my eyes and knew I was nearly home, I can see the spot as clearly as if though it was yesterday. It was just a flash lasting a few seconds but before and after that moment, I remember nothing. The second thing I remember was that I was in a different house. I was in bed and staring at the pattern on the wallpaper and on the mantle shelf was a clock, which had a six-sided frame and a small face. I always seemed to focus on the same things night after night.

Another early memory was hearing a gunshot and a short time later my father came in with a dead barn owl It was later stuffed and remained an ornament in the house for many years.

I cannot remember when we moved to the new house[1] but I would be about three years old at the time. My brother John was certainly there too because one day he was crawling over a steep bank at the side of the house when he fell over and badly gashed his head on a rock. I suppose a doctor was called and later when John recovered a wee white mark remained to show how he narrowly missed death.

I have a dim recollection that from time to time I went to my grandmother's house. Sometimes I would be taken there but latterly I remember slipping away when I was sure of the road. It was only half a mile distant and there were no motorcars in those days and few people moving about and only an old cart or a cab on the road.

As the first grandchild, I suppose I was fussed over as is usual in such cases. I remember my old grandmother sitting in an armchair beside the fire with a mutch on her head looking very old and frail. On occasions, she would be taken to the door where she would sit for a while, but I never saw her walking. In later years I was told many stories about her young days. She came from Carrick Castle on Loch Goil where her parents had a farm called Cuilimuich meaning in the Gaelic "the place of the pigs". She was born in 1808 and as she grew up she was given the task of taking the surplus farm produce to market. Transport in those far away days was by boat, or horse and cart. Otherwise one had to walk and time was less important. In isolated places like Carrick Castle it was not easy to sell farm produce but a new market was developing in Helensburgh and it was there she had to go. So, early every Friday morning she set off in a rowing boat for Portincaple on the east shore of Loch Long with the baskets of eggs and butter. Leaving the boat she had to climb up the rough track to Whistlefield and down to Garelochhead, from where a carrier's cart took the produce on to Helensburgh. At that time Garelochhead was a collection of crofts, and traces can be seen still of the remains of these. They would all join up on market days and most of them would have to walk behind the cart to Helensburgh. On the return journey the baskets would be used to take home the luxuries of tea and sugar and other produce. For my grandmother, the journey home over Loch Long, past the Dog Rock and Corran Bay could be good or bad depending on the weather. Sometimes she could not get home owing to storms. On one occasion she had to shelter under a rock all night because of a dense mist on the tracks over to Loch Long. It was on one of the market days that she met my grandfather as she passed by his joiner's workshop. He had come to Garelochhead from Liverpool to

---

1   Possibly Daisy Bank, more or less opposite the Cottages.

start a business. He was a native of Ayrshire where his father was a dry-stone dyke builder. As new houses were in demand he was just beginning to build up a business. I never saw my grandfather, as he died two years before I was born. I was told that he was a good flute player when I enquired about the flutes I saw lying here and there.

The joiner's shop was across the road from Oakfield Cottages and it was quite a big place with four double benches. When my grandfather died, my father took over the business and paid a rent to my grandmother. I spent a lot of time in the shop and the smell of shavings and sawdust soon got into my system. To me it is the most satisfying of all smells although I ceased to work with wood as a craft fifty years ago.

On the south side of the shop was a place always known as the saw-pit. For a long time I could not understand what a saw-pit was but later learned that it was the place where the great logs were sawn by hand into boards. The logs came to Port Glasgow by boat from America and after seasoning there, were bought and towed as a raft of logs to the head of the loch at Fernicarry. They were beached and then placed on bogies or most probably on sledges and pulled to the sawpit by horses. Once there they were stacked and when required were levered on to a special frame-work where two men, one above and one below in the pit, proceeded to cut them into planks of the required thickness. These planks usually of first grade yellow pine or pitch pine were used for all sorts of jobs such as panelling, doors, windows etc. Sometimes a second or third grade was cut for roofing etc. The saws used must have been very sharp and made of the finest steel, but it must have been heavy work and I used to think of the toil and sweat involved. I have often seen these planks in the better class houses and I have in my own house roof spars of the finest yellow pine and doors of the same material well over a hundred years old, as sound as on the day they were erected. The hand sawing of the logs however ceased long before my time and the only remaining evidence of the practice that I know about is the empty pens in the Clyde above Port Glasgow where logs were seasoned.

There was a wood rack nearby for the rougher timber like roof spars but all the planed and better class timber was placed under the workshop where there was ample room. The timber would have spoiled in rain and sun and such things as flooring and sarking had to be protected. All this timber had come from a sawmill in Glasgow and was replenished as required.

I got to know, by stages no doubt, the surroundings of the cottages but exactly how I passed the time there I do not know. The family at home was increasing and I soon had another brother and sister and in fact we had moved to a larger house in a more central position. After a year we again moved to a still larger house, which was available. Time was passing and the day when I had to go to school was near.

Just about this time I had been taken to live temporarily at my Grandmother's home and when the day arrived for me to go to school, which I remember well, my Aunt Maggie put a school bag over my shoulder and took me outside. She had seen a man with a cart and horse passing by, going to the nearby smiddy and had asked him if on his way back he would take me as near to the school as he could. So he stopped and I was hoisted into the cart and sat beside him on a plank. I can see it all again and I was not very happy as we moved away. In less than ten minutes I was lifted out of the cart close to the school. But the sight of the children running about must have scared me because with little hesitation I set off for home. My Aunt Maggie lost no time in taking me by the hand and it was not long before I was sitting at a desk in the school where I was to spend the next nine or ten of the happiest years of my life.

I remember little or nothing of my first few years but in my mind's eye I can still see Miss Wood the teacher, a big kindly woman, with the coloured balls on wires on a frame standing beside the blackboard. Those were golden years and there was plenty of time to play with other boys and join in their pursuits. Every morning during the school days a large number of children passed by the cottage on their way to school. They came from Finnart and Ardarroch, Portincaple and Whistlefield and some of them had to walk about three miles each way. I usually joined some of them. They got their stockings and feet wet at times but good fires were provided at school, which dried their stockings at least, before they left for home in the afternoon. They seemed to thrive on the fresh air and exercise.

After a time I returned to live with my parents and I found that I had now two brothers and two sisters but I regularly visited my grandmother and aunts. I remember one incident in the great frost of 1895. My father was taking me home and we were walking up the place called Browns Brae in what, to me, seemed deep snow. It was my first time among snow and I had some difficulty in walking. In later years my father told me how the snow came just before Christmas and after a few days a thaw set in. Early in the new year heavy frost came and, without a break, continued until early in March. Even Loch Lomond was frozen solid and horses and carts were driven over it. No such severe frosts have before or since been known round the Gare Loch.

When I visited my aunts I usually paid a visit to the joiner's shop maybe just to stand and look and watch the men planing and sawing or making windows and doors. Occasionally I would glance at an empty bench on which stood a few beehives and as I grew older I used to gaze at them in wonder but all I could learn was that they had been made by my deceased Uncle Robert. He had been an expert beekeeper and had kept up to date in all the developments of bee keeping that were taking place. I learned later that these hives were models of the first movable comb hives used in this country and when I began to keep bees myself I made good use of these hives.

My aunts told me all the usual fairy tales when I was young and I believe the little folk were real to me. There was a round conical hill to the north-east which I was told was the Fairy Hill and I could see it clearly from outside the front door, and there, fairies were supposed to live. I often wished I could go and see them but even in later years when I could have gone I thought it too isolated and far away. There were rhymes, which I learned by heart about fairies and everyday events. On Saturdays my father used to recite:

'This is wee Saturday, the morn's Cockielaw, we'll all rise on Monday and gie the wheel a ca.'

I am still wondering what Cockielaw was (it was Sunday) but where did the word come from and how should I spell it? When it rained and I was confined to the house, watching the rain pattering on the road, my Aunt Maggie would say:

"Rainy rainy rattlestone don't rain on me. Rain on Johnny Groat's house far across the sea",

and I would wonder who Johnny Groat was.

Riddles of course were numerous but now I can remember only two:

"Through the woods, and through the woods, and through the woods it ran, but though it was a wee thing it killed a muckle man", The answer was a bullet.

And the other one was,

"Come a riddle come a raddle, come a rote tote tote, a wee wee man in a red red coat, a staff in his hand and a stone in his throat".

Maybe there was another line or two, but anyway the answer to the riddle was a cherry.

Another ploy, which my aunts used when I asked too many questions, was "going to market". They would sit on a chair and place one leg over the other. Then they grasped my two hands while I sat on the outstretched leg. They would then raise me up and down to the following refrain *"This is the way the wee boys ride, the wee boys ride, the wee boys ride. This is the way the wee boys ride when they go to market"*. I could never get enough of this until the time came when I got too heavy for their old legs.

In April 1897 my old grandmother died and my brother John and I went to her funeral. I remember looking out of the cab window on our journey to Rhu and looking toward Rahane and Clynder and thinking how bare and bleak the countryside looked. About ten days later my Uncle John died and again we were on our way to Rhu churchyard. A few weeks earlier, my brother and I had been taken by my uncle to Greenfield Farm and on the way, we saw lots of lapwings or pewits. On our return he took us in his boat across the loch, where he had his moorings.

At this time I was seven years old and growing quickly and doing well in the infant class at school. Sometime in May of the same year an event of great importance took place locally and nationally. It was the Diamond Jubilee of Queen Victoria. Preparations for the event had been going on for some time and a bonfire had been prepared. There was also a torchlight procession. The school also made preparations. All of us were marshalled and marched through the village to a place not far from the pier. Once there we saw a new memorial drinking fountain, which was formally opened. Then we were each given an inscribed tin mug and treated to tea and buns. However, the thing that made the most impression on me was the torchlight procession. I do not know why I was allowed to stay up so late. I suppose it would be getting on for eleven o'clock. We lived in a flat[2] at that time which faced the Church and the road.

I could see the lights of the torches as they approached and I got a fine view of the procession as it passed by on the road, returning by the Shore Road. Doubtless I was in bed and asleep before it finished but the flaming torches long remained in my memory. It was indeed a red-letter day. Most houses had a picture of the old Queen hung in a prominent place.

---

2   In Glencairn Terrace

Sketch of Diamond Jubilee Fountain

2. The Blacksmith; Work and hobbies etc.

Sometime after my grandmother died I was again transferred to the home of my aunts. I suppose the reason was that my mother had too much to do with the increasing family and this time I was to stay for a few years. I was delighted as I was in the place I knew best. Lots of horses passed by the cottages every working day, going to the smiddy and I think by this time I was feeling bolder and would walk up to the smiddy, which was about a hundred yards up the road, and stand at a respectful distance observing what was going on. In the early morning, before I got out of bed, I would hear the clang of a hammer on the anvil.

Sandy Gilmour was the blacksmith and his assistant was Hugh Buchanan who had been with him for many years. Strange to say the first smiddy had been the cottage in which I was born and Sandy's father was the occupant. My grandfather had bought the property from him and the blacksmith had built a new house and smiddy on a large piece of land adjoining the former feu.

Sandy was a very quiet man and I was shy and I don't remember passing a word with him at that time. He always kept three or four cows, dozens of hens and ducks and a few turkeys, all of which were housed in a building attached to the back of the smiddy. He had also several pigeons of various kinds. He had a large garden, which provided him with ample produce and with cows and hens he had plenty of manure for the garden.

A good distance away from the animals he had his apiary with up to twenty hives in it. Actually, I never got near enough to the bees to see them working nor had the courage to ask Sandy if I could look on when he was working with the bees. I don't suppose he ever thought a little boy would wish to look into a beehive. In fact most if not all the information about the bees came from his assistant Hugh Buchanan.

Sandy had a pug dog and another small dog and every day after his meals he would take the dogs for a walk up the bank of the burn or perhaps he would take them up past the wee wood on the road to Fernbreck or in the evening he would saunter up the hill to have a crack with the forester with whom he was very friendly. Sandy lived with his two sisters one of whom was married to a sea captain who seldom came home, and she had a parrot and a cockatoo, no doubt bought by her husband. The birds were put outside in their cages in good weather and I used to hear the parrot talking away but I was never near enough to hear what it said. All the fowls had free range in the wood above the smiddy and many of them laid eggs there. Sandy had grey hair and a dark complexion and always wore a skipped bonnet similar to what fishermen wear.

As I watched the burning sparks flying from the red-hot iron as he shaped it, or heated the tools and sharpened them, or ringed the big farm wheels with their bands of iron, I did not ever dream that before many years had passed all would have vanished forever. The quiet country road where horses pulled the loads up the steep hill would echo to the roar and fumes of the petrol wagon and motor car and the deserted smiddy would be falling into decay.

One day when I would be about ten years old Sandy passed by with a fork in his hand and my aunt told me that he would be on the way to the shore to dig up worms for fishing. I was the proud possessor of a new fishing line with two hooks and lead sinker attached. I suppose I bought it in my aunt's shop for a penny or two. My aunts Jessie and Mary had for many years had a draper's shop in the village where they sold toys and odds and ends in addition to the usual drapery found in most country shops at that time. During the summer there were lots of whiting in the loch and Sandy usually fished on Saturday afternoons by himself so when he came back from the shore my Aunt Maggie walked out and asked him if he would take wee Willie with him and he readily agreed.

After dinner he came down the road and with never a word I joined him. Sandy had his rowing boat moored on the west side of the loch and in a short time we were there and the boat was pulled in and bailed dry. I got into the boat with my precious fishing line and sat in the stern seat. I had no idea where we were going and all I remember was that Sandy headed down the Loch and after a while we passed Mambeg pier.

A little further on there was a good place for fishing as I learned later and so, when we got to the place Sandy dropped the anchor and we were ready to fish. I had never seen such big worms as those coiled in a tin. They were monsters to my eyes. Up to a foot long, they were reddish in colour with numerous spikes on both sides, and a mouth that could open and give one a nip. They were called by the locals "ruchies" or rough worms and fish liked them better than mussels. I suppose they swam about in shallow water when the tide covered the sands and the only time they could be dug up was when the tide was well out. I have no doubt that Sandy put a worm on each hook before I could fish. In a short time a fish gave my line a tug and I instinctively tugged in return but that was the end. It might have been a big fish, I don't know, but my line had broken. I suppose there were tears in my eyes, Sandy took a spare line out of his basket and after baiting it said, "There you are Willie, fish away." That was all he said, I don't remember if we caught any fish and I don't remember if he said another word, though I expect he did, but I look back on that episode with not a little nostalgia.

What I did not know at the time was that Sandy and my two deceased uncles were great friends, living as they did within a hundred yards of each other and were probably at school together. The craft of joiner and carpenter linked up with the work of the smith in the making of wheels for carts and barrows, and bolts and a host of other fittings used by joiners.

One more thing was that he liked to fish off the rocks over the hill in Loch Long. Hugh, his assistant, told me that he once caught a salmon on one of these trips and when I was old enough I never missed the chance of fishing off the same rocks for lythe, cod and saithe.

The Smiddy

3. School Dress;
Sunday school; Barefoot Boy etc.

The normal dress for most boys at school was a Norfolk jacket, an Eton collar and shorts with long stockings, but from May onwards, after school and on Saturdays, the majority of boys took off their shoes and stockings and went about in their bare feet.

Being a seaside place with plenty of boats, the shore was a haunt of most of us. And there were plenty of meadows and burns about. Furthermore, it economised on our footwear which in large families was a big item of expenditure. I suppose I would be ten years old or a little more when I first started going barefooted and I looked forward eagerly to getting permission from my mother to be a barefoot boy. What a delight it was to have the feel of the road and the cool grass under foot. The roads in those days were stone and rough, very unlike the smooth roads of today but they were dusty and kindly to the feet and from June to the end of September were warm and we took care to avoid the loose stones. In a few days we got used to it as the skin on the feet hardened and toughened. Sometimes I would get a cut from a broken bottle and the remedy was soap and some sugar on a bandage and my shoes on for a few days and then barefoot again.

One of my favourite poems is "The Barefoot Boy" by Whittier. Here are a few extracts and Whittier describes it perfectly and I live again the joys of yore:

Blessings on thee, little man.
Barefoot boy with check of tan!
With thy turned-up pantaloons,
And thy merry whistled tunes;

With the sunshine on thy face,
Through the torn brims jaunty grace;
From my heart I give thee joy,
I was once a barefoot boy.

Prince thou art, - the grown-up man
Only is republican
Let the million dollared ride!
Barefoot, trudging at his side
Thou hast more than he can buy
In the reach of ear and eye.

Oh for boyhood's time of June
Crowding years in one brief moon.

Larger grew my riches to
All the world I saw or knew,
Seemed a complex Chinese toy
Fashioned for a barefoot boy.

Cheerily, then my little man
Live and laugh, as boyhood can.

All too soon these feet must hide
In the prison cells of pride
Lose the freedom of the sod
Like a colt for work be shod.

Ah! That thou couldst know thy joy,
Ere it passes, barefoot boy.

There was only one day in the week when bare feet were not allowed and that was Sunday. I had to dress up for that day and put on my best clothes in order to go to Church and Sunday school. The Church service did not start until noon and therefore in spring and summer it was usual to take a short walk along the burn and gather flowers before preparing for Church. All play was frowned on but one could walk in the fields or, walk along the burn observing wildlife or, sit down and read a book.

The Sunday school lasted until two o'clock and after lunch I would take a walk with my father. There was Church again at six o'clock but I don't think I went in the evenings until I was older.

As far back as I can remember there was always a week or two of haymaking in the Meadows. The smaller meadow was rented by the football club but the larger meadow across the burn was always reserved for hay by the farmer. The hay was seldom cut before the second week in July and sometimes later. One day two or three farm workers would arrive from the farmhouse, a mile and a half away, with two splendid Clydesdale horses drawing the mowing machine and, crossing the bridge over the burn, would begin operations.

As swathe after swathe fell to the machine I would watch the dogs working round the field waiting for the rabbits that were hiding amongst the hay. A day or two later, depending on the weather, the swathes would be turned over to allow the sun to dry the hay. Then next day or as the hay was reasonably dry the big day arrived. Every woman and child who could help came to the field and the summer visitors were numerous in the cottages.

Old Greenfield was always present in my early years. I don't think he did much work but somehow added to the importance of the occasion. His name was Duncan MacFarlane but to every native he was "Greenfield", and how that came about is not easy to explain. However, he had a very large farm, of a thousand acres or more, most of which was hill land and only fit for sheep grazing, and it seemed to be the rule in Scotland at that time to call a man who had a big farm by the name of his farm. Not far away was another large farmer who was always called "Faslane" after his farm and another in Glen Fruin who was called "Strone" after his farm. Smaller farmers, in contrast, were always known by their surname.

Sam was the ploughman who directed operations. The first operation was raking the hay into lines with a big broad machine with curved prongs that could be raised and lowered to catch and release the loose hay. As soon as the raking started we set about making small heaps of the hay sometimes called cocks, but in Scotland usually called coils. The whole field would thus have the hay in coils. The following day, if the weather was fair, the coils would be amalgamated into small stacks or, they might be loaded into a cart and taken to a central site where the large stacks would be made. It all depended on the state of the weather, the main thing being to protect the hay from the rain. In this case, as the hay had to be eventually carted to the Home Farm they had a machine with very low wheels on to which a good-sized stack could be pulled with the minimum of effort. There was a good deal of raking to be done after the machines had done their work and everyone seemed to be kept busy until the last of the hay had been stacked. When the hay had been cut the plovers and corncrakes in the meadow disappeared.

The corncrakes had come in the late spring just as the grass was growing fast and one day we would hear the sound of that mysterious bird. As soon as dawn broke, about three or four o'clock, one could hear the 'creck, creck, creck, creck, creck'. My Aunt Jessie, the oldest of my aunts told me she could not sleep after that time. It kept on at intervals throughout the day uttering its strange harsh cry. One day as I was walking along the burn side after picking a few flowers and looking for birds and nests, I heard a corncrake some twenty or thirty yards away. I had my feet bare and was walking noiselessly. I advanced slowly and then stopped until I heard it again when I advanced a little and then stopped. I made up my mind as I was getting close to the bird I would rush it as soon as it cried again as I knew it would quickly run away among the grass once it knew I was there. So at the very first crake I dashed forward and there at my feet rose the corncrake, flying low and rapidly over the grass. It was brown in colour and similar in size to its relation the water rail. I had seen my first, and last, corncrake and I was happy.

Less than a hundred yards down the road from the cottages there was a wood, which ran along the Rosneath road for about four hundred yards and uphill for about the same distance. My aunts always called it the Briery Wood although there was not a brier in it. But when they were young the previous wood had been cut down and briers had grown where the trees had stood. In the course of time the native oaks had taken hold and now it was a wood composed of those thin oaks, about thirty or more feet high.

It was the favourite wood for collecting sticks for firewood and two or three days a week I would help in taking home bundles of rotten branches. And there was an abundance of leaf mould for the garden. It was a pleasant wood to walk about in with birds and squirrels all around.

At the corner nearest to the cottage there stood a majestic oak tree known as the "Bogle Tree". It must have been hundreds of years old and I never knew of any oak tree of such size in the district. The reason why it was called the "Bogle Tree" was that there was a hole in the trunk about ground level and into this hole the Bogles came and went. That is what I was told by my aunts. Nobody could tell me what a Bogle was like and although I have since read stories about Bogles I am not any wiser. At the time I thought it was some kind of animal with a human like head and it only came out at night when it could not be seen and I never liked passing by alone in the darkness. There is a story attached to the tree which may be of interest:

One day a number of men arrived at the site and although I was young at the time and had been going to school for four or five years, I learned that the men were foresters and had come from Rosneath. When I came home from school later the mighty tree was lying on the ground and the foresters were lopping the branches. The following day they arrived with a special wagon designed for moving heavy tree trunks and the trunk of the Bogle Tree was, after a lot of trouble, on its way to Rosneath. It appears that there had been a long-standing controversy between Sir James Colquhoun of Luss and the Marquis of Lorne, later the Duke of Argyll, the owners of the adjoining lands, about the ownership of the tree. The Bogle Tree stood on the boundary between the two estates. It had a firm base on Argyll land but the roots of the tree spread over the bank and beyond what had once been a wall to the edge of the public road. I suppose that each proprietor owned part but not the entire tree. It was a pity that the Marquis of Lorne had so little to think about in ordering the tree to be felled, or it might yet be standing, the monarch of the wood. After the tree had been felled I looked at the hole in which I used to think the Bogles lived and found it was only a few inches deep, but at the time it was very real to me.

The Bogle Tree

A few yards away to the north of the Bogle Tree the boundary wall turned at right angles and went up hill following the contour of the country until it finished two miles away on the shores of Loch Long. When I was young I did not take much notice of it, it was just a wall to me. I seldom or never clambered over it. This arch dyke was built of dry stone and in later years I took more notice of it particularly when searching for white heather in August. As far as I know it was built towards the latter part of the 18th century probably about 1780. At the base it is about three feet broad tapering to about fifteen inches at the top and from six to seven feet high. Every stone is fitted carefully, with headers and projecting leaders every thirty inches to stabilise it. The stone is of hard whinstone and obviously was quarried at intervals along its length. On the Loch Long side of the hill it runs sharply down hill to the Loch and I often marvelled how the builders got the stones down towards Loch Long. The only practical way was to use horses and sledges. The men who built the wall were undoubtedly the elite of their craft and the wall stands today as it has stood for two hundred years a monument to their skill. Over the years I have observed stone walls in many parts and have not yet seen one even faintly resembling this work of art.

Not far from where the Bogle Tree used to be there stands a peculiar tree, known to me as the "Four Trees". It is an old gnarled oak that has four trunks, each about two feet in diameter. It may have had, when young, four shoots from one sapling or four saplings fused into one tree. Anyhow it is now about five feet in diameter with a slight hollow in the centre and is about forty feet high. When I last saw it a year or two ago it looked to me just as it was when I was a boy.[3] It grows on a knowe at the side of the road to Rosneath and the knowe is a vantage point for looking down the Gareloch as far as the narrows at Rosneath. I used to play in and around it and when asked where I had been I would reply that I had been to the Four Trees or the Knowe.

---

3   The 'Four Trees' or 'Four Oak Trees' are still there and have changed little.

A recent photo of the Four Oak Trees

4. Hogmanay; Sports; Cow & Hens etc.

On Hogmanay for some years, just before midnight, my Aunt Maggie would take me down to the Knowe to hear the New Year brought in.

The Parish Church bell would be the first to ring closely followed by the bells of the steamers lying at the pier. Then shots would be fired here and there and the grand climax was when Sandy the blacksmith fired one or two blank shots from an old cannon which no doubt could be heard for miles, and then back home to a glass of ginger wine and a piece of cake or shortbread and then to bed for tomorrow was the big day of the year.

January the first was the day when old friends met, and some came long distances. Unless it was a Sunday, the first of January was the day when the annual sports meeting was held and it did not matter whether it was frost or snow or just the usual good weather.

For long years the sports had been held on the football field just in front of the cottages and the day before preparations were made by a band of volunteers. The field, of about three acres, was flat and ideal for the purpose. Posts were erected and wire or rope strung round to provide the sports arena. At noon the sports began and continued for probably three hours. First of all the children had their turn with races for different ages with three leg races to cause some fun and the sack race. Then there was a biscuit eating competition. I liked that probably better than anything else. Each child was made to stand in a line and was given a butter biscuit to eat. Now these biscuits were no small affairs like the ones you buy in a bakers today. The name butter biscuit was given because one needed plenty of butter to swallow the pieces you had bitten off, and believe me these biscuits were dry.

The children competing were anything from ten to fifteen years old. At the word "GO"! they set about the biscuits and no comic turn on the stage ever caused so much merriment and hilarity as the antics of the children trying to swallow these biscuits. Each was determined to win a prize and three shillings would be a fortune to most of them but no bits were allowed to fall to the ground. Every crumb had to be eaten. I think I tried it once but that was enough. Another competition that I also liked was called "cuddies". A small boy got on the back of a big boy and the object was to drag another small boy off the big boy. I won this competition once, my cuddy being Johnnie McKichan of Mambeg. I do not remember if I ever won the egg and spoon race, but I tried that too.

The men's sports followed and were of the usual pattern of sprints, quarter and half mile races, married men's race, putting the stone and tossing the caber. Then there was the long jump, hop step and jump and the high jump. Finally there was the "Cutting of the Ham" competition. A fine Belfast Ham was hung on a bar. You stood with your back to the ham, were blindfolded and were marched twelve paces. Then you had to turn around three times and with a cane in your hand had to march back to where you thought the ham was hanging and strike it with the cane. The one who found the ham with the first strike was the winner, and in the case of a tie, which was a rare happening, a deciding round was necessary to award the ham.

There was always a piper at the games, usually a West Highlander from the steamer, and he played everything he knew. Almost everyone in the Village turned out to see the games even if only for a short time and it was all handshakes and good wishes wherever you went. One person however saw little or nothing of the Games. She was sitting on a stool at the entrance to the field and was selling Spanish oranges presented to her by the Games committee as a New Year gift. She was Susie Reid one of the characters of our village. She came from Portincaple over the hill and lived with her husband Jimmy and her cats. Her home was an old upturned fishing smack with curtains on the windows and everything spotlessly clean. Almost every day except Sunday, Susie came over the hill carrying a basket in which there might be a few fish or something else she could sell to the villagers.

Susie's Castle

Susie knew everybody and everybody knew Susie. She talked to anybody and everybody and was obviously an intelligent person but had two vices; she liked a dram and she smoked. It was on her way home that I saw most of her, by then she would have had a few drinks and odds and ends in her basket. It was not easy to tell her age but her legs were obviously feeling the strain and her first halt was at the Foresters Bridge a couple of hundred yards beyond the Smiddy. There she would rest on the stone parapet puffing away at her clay pipe. Then on again for a bit, and another smoke and I have no doubt she eventually reached Katie Munro's pub at Whistlefield. Bear in mind that in those days one could get a half of whisky for two and a halfpence. Thus fortified she toddled on, downhill now, to her home. Picture post cards of Susie and the upturned boat were best sellers to visitors. She was a kindly person and so far as I know never said an unkind word to anyone.

There were two cats in my grandmother's house. One was called Greynx and the other Tumphy. The grey one was very old I think and it did not welcome attention from me. It would spit at me and show its claws and I confess I did not like it. But the other was quite the reverse. It was a true tortoiseshell. It was a kindly cat and never scratched me. Tumphy was a great hunter and anything that moved was her prey. She often came in with young rabbits and one occasion I got my aunt to put the rabbit in a press[4] with the intention of giving it freedom. But later on when I opened the press the rabbit was gone.

Another morning when having my breakfast Tumphy came in and laid a slow worm in front of the fire. I was terrified. Later on I found that slowworms were quite common and on occasions found some near the house. Slowworms, although they resemble snakes are quite harmless. Unlike snakes they have no markings on the body but are slate blue in colour. Even afterwards I was frightened of snakes and I remember seeing an adder at the top of the Briery wood when having a stroll on a Sunday afternoon with one of my aunts. Believe me I ran and so did she. The only other snake I remember was on one occasion when fishing at the Lochan, which supplied water to Cove and Kilcreggan. I found a dead Ringed snake[5] on the bank. It would be about two feet long and on reading a reference book found they were harmless.

---

4   cupboard
5   More often known as a grass snake but these are not native to Scotland. It may have been an adder.

My Aunt Maggie the youngest of my aunts had kept a cow for many years. It was kept in a byre below an old house that was attached to the joiner's shop. This derelict house was referred to as "Nannies" and was used by my father as a store and lumber room. Nannie I learned was an old, lonely woman who had been given shelter there by my grandparents long years before. The byre had one stall and the remainder of the space was occupied by the hens, their perches being separated from the cow by a half partition.

There was a small window that gave a little light when the door was closed. These hens never stopped laying eggs summer or winter. Through the autumn and winter, when eggs were scarce elsewhere, they were plentiful here. I was told that the reason was the warmth emanating from the cow. The hens of course were of good strain, Black Leghorns, White Leghorns and best of all Black Minorcas. The eggs of the latter were large but all the eggs had a fine flavour as the hens had free range.

In later years when on a visit to my aunt I invariably had my tea and taking pride of place was a boiled egg. What eggs! It is many long years since I have tasted eggs like those I got from my aunt. The white was finely granular and the flavour I cannot describe in words. I suppose there must have been something in the soil and the good feeding they got, for they were always superior to eggs I had elsewhere. Then there were Abemethy biscuits the bakers made and fresh butter and carvey on top. Carvey is seldom seen nowadays but at that time it could be obtained everywhere. For those who have never seen or tasted it, it was simply sugared caraway seeds, in appearance much like grains of rice. Then there were home baked soda scones, pancakes, water scones, and potato scones.

The cow was milked in the morning and when it came home from the hill in the late afternoon, I often joined my aunt at milking time and would watch the rich creamy milk being drawn from the cow's udder and listen as the milk swished into the pail. The milk would then be taken to the house and poured into the large flat pans to cool.

Often there was a surplus of milk and it had to be turned into butter. The long upright churn would be placed in the kitchen and the paddle would be worked up and down. Slowly the granules of butter would appear and eventually the rich gleaming butter would be moulded in the wooden moulds ready for using or selling. And the buttermilk or sour milk as we called it would be used for the morning porridge. At that time a cart came all the way from Craigendoran selling sour milk and butter and no one needed to be short of either.

The cow grazed on the hills to the north and every morning after breakfast the cow was let out of the byre. It went through a narrow passage to reach the road and would make its own way up the brae to the hill. Usually the blacksmith's cows would be let out at the same time and they would all go together and keep together all day. They knew when to come home in the late afternoon but occasionally they had to be brought home when they had gone too far in search of sweet grass among the heather, bog myrtle, and brackens. As I grew older I was often given the task of searching for them and as soon as I got up to the forester's house I would start calling "Proochy Lady", "Proochy Lady". That was what my aunt always cried when she had to call them. I had no idea what the words meant but the cow certainly knew the voice and responded. Every cow she had had was called "Cherry" and even when I tried that there was little success. I never had the courage to ask my aunt what "Proochy Lady" meant but an old farmer near Oban told me that it meant "Come here" in the Gaelic. That puts me in mind that my grandmother was a fluent Gaelic speaker and my aunts knew quite a bit of the language although they never used it, except my Aunt Mary who occasionally dropped a word or two.

There were other cows too on the hill which came from Fernbreck, but they seldom mixed with the cows from our place. The land belonged to the farmer who in this case was Faslane and he ran sheep mainly but always had a herd of West Highland bullocks which he fattened up yearly before sending them to market. I remember one occasion when during a hot dry spell, and clegs and flies were very prevalent, these bullocks stampeded. At the time I had just left the highroad and was walking along a fence which divided the meadowland from the hill land, when suddenly, I saw in front of me a large herd of West Highlanders with their massive horns making straight for me. There was only one thing to do and I got through that fence like an eel and dived among the rushes with which the field was covered. As the thunder of hooves and the snorts of the cattle passed I buried my head and held my breath. I was too scared to lift my head and all the time I wondered if I was on the right side of the fence or whether the mad beasts had broken down the fence. But the fence was made of strong posts and wire and they were making for a wide gap in the hedges which bordered the road, through which I had entered a few minutes earlier. How long I lay there I do not know but when all was quiet I got up and keeping a close watch for the cattle I made my way down the road. Then, suddenly I saw them in the wood just above the road. And behind them and driving then back with heavy sticks were the two blacksmiths from the smiddy. I gathered later that when the cattle came galloping down the smiddy brae the smiths rushed out and turned them. I don't think I told anyone of my experience but often I thought about it and wondered if I had been a coward, and then again I would see these twenty odd cattle with their great horns charging in a body and only a small boy in their path. And although I used to laugh after dangers were past, in the years to come I must confess that I never laughed at that one at any time.

The cows had a very large grazing area. The farmer charged four pounds a year for the privilege of grazing and it was a very reasonable charge for such rich feeding. I sometimes would see him coming to the door to collect his rent. Faslane was a fine upstanding man and like all the other big farmers in the district he was a MacFarlane.

There was always a calf in the spring and it was kept in a pen below the joiner's shop and fed with the cow's milk, and I remember the milk was not used for a week or two for domestic purposes as it was called Beastie milk. There was always a stack of hay for the cow and plenty of turnips and meal. The cow and the hens kept my aunt quite busy. She had a fine garden on the low side of the road and this was in main a flower garden with in addition a potato plot and some bush fruit. There was an abundance of manure from the cow and hens and the garden was in splendid condition as a result. The soil was dark and deep and everything grew with great vigour and there were always one or two Scotch thistles, which grew to six or more feet high. They were beautiful plants but I never see them nowadays. My Aunt Maggie, until she was over seventy, dug and planted the garden all by herself. Living in a village with many fine gardens she was always being offered bits of plants which she did not already possess. Every corner had something of interest. When the visitors came and especially day-trippers she used to put together large and small bunches of flowers and sell them at the door and there would be a sprig or two of southernwood or mint in each bunch.

The fronts of the Cottages were ablaze with flowers, Roses, Clematis, and Kerria Japonica covered the walls and porches. In the borders herbaceous plants and Nasturtiums made a blaze of colours and the window boxes filled with Begonias and Forget-me-not. White chuckies edged the borders, Musk and Mignonette sprawling over them and no picture post cards of these cottages ever conveyed the beauty of their appeal, which made passers-by stop and admire. Even in the dull winter months there would be primroses in full bloom with tufts of crocus to keep them company.

Oakfield Cottages

5. The new road to Arrochar; Reliability Trial etc.

Our cottages stood at the bottom of Whistlefield Brae on the road to Arrochar. It is a steep hill most of the way and a heavy hill for any horse to pull a load and usually a trace horse had to be used to assist. The only horses that made light of it were those belonging to Colonel Marryat of Finnart House. They were Belgians and actually would run up the Brae with a carriage if not held in check. At the top of the hill was a place called Tim-na-Cross where the path from Portincaple to Garelochhead crossed the road before Whistlefield or the present road existed. It was shorter and continued in a more or less direct line to Garelochhead coming out at where the bowling green now is. Only faint traces remained in my youth. The road or path to Portincaple can still be traced from Tim-na-Cross and although I have walked it many times in parts it is obliterated with peat cuttings and stagnant pools of water.

The road from Rosneath to Arrochar was built by John Duke of Argyll and there is[6] an inscribed stone almost directly opposite the gates to Ardarroch House which reads as follows:

This Road was made from
The Castle of Rosneth
To Tenne Cloich
In the year 1777 by
His Grace John Duke of Argyll

Erected by Donald Fraser

Tenne Cloich is the name of an old inn which once stood at the cross roads in Arrochar near where the Hotel now stands and means in Gaelic "The house by the big stone". The road was made by the Duke to give direct communication by road between Inveraray Castle and the old Rosneath Castle. Strange to say the entire road from our cottages to Arrochar was built on land belonging to the Colquhouns of Luss and I have no doubt no objections came from the Colquhouns. It may be of

---

6   The stone is still visible but the wording is now difficult to read

interest here to say that the Argylls first came to Rosneath in 1489 in the time of the first Earl of Argyll.

Ardarroch and Finart are the two mansions that stand opposite each other where the road joins Loch Long about three miles from Garelochhead. They had beautiful grounds and employed a large number of both outdoor and indoor staff. R. Brooman White was the owner of Ardarroch and spent at least half of each year there with his wife and family. He was a wealthy man and his hobbies were shooting, motoring and orchid growing. He had a staff of four for the orchids and many more for the outside work. Sometimes I had the opportunity to peep into the orchid houses and see some of the magnificent blooms many of which went to London. He was the first in the district to own a motor car. Among the first I remember seeing was a big red Mercedes of sixty horse power and it used to race up Whistlefield Hill like an express train making a noise like a machine gun in action. I don't know how many tyres that car used up in a year for in those days the roads were covered with loose stones, but it could be no small number. He had one of the first Argyll cars and then a big Daimler which he had for some years. The Whites were, I suppose, aristocrats and the family was brought up as such and had little to do with the village folk.

The owner of Finnart was Colonel Marryat, late of the Manchester Regiment. He lived there with his wife who was a sister of Sir James Caird, the jute magnate of Dundee. They were a quiet country loving couple and lived there all the time. They had a fairly big staff and in their case they had a grieve in addition to a coachman and a number of gardeners. With some fine specimen trees and perfect lawns Finnart was a delightful place to stroll about in.

The owners were lovers of all kinds of animals and the animals knew it. They were, I believe, the people who first introduced the grey squirrel to Scotland and more than once I saw Mrs. Marryat feeding them. I never at any time heard anyone say that they did damage to trees or crops and I remember being told by a head gamekeeper near Helmsley in Yorkshire that they did no damage as far as he knew. There were plenty of red squirrels in the district at that time and in fact they were predominant and I used to see their nests or dreys in fir trees. As a matter of fact I delighted in chasing them through the woods watching them jump from tree to tree.

The Colonel and his wife had a brougham and a dogcart and they had very fine horses. As I have already stated they were Belgian horses of a fawn colour. I don't know how many they kept but they were full of fire. The steep hill over to Whistlefield would try out any horse but as they passed by on their way to the village they had not turned a hair. There were some fine horses about in those days.

In course of time Mrs. Marryat was presented with a fine car by her brother Mr. James Caird. I must confess I have no idea what make it was but I certainly remember its colour. It was bright daffodil yellow and the coachman became the first driver and he certainly had some difficulty in changing gears at the foot of the hill. But it was a beautiful car and caught the eye wherever it went. Some years later Sir James died and Mrs. Marryat became a very rich woman but as far as I know she gave most of her wealth away to finance scholarships for music, the Caird Hall in Dundee, and many other benefactions. Now Ardarroch and Finart are shorn of their beauty; the owners are no more, and nothing but oil tanks among the remains of the former fine buildings and perfect lawns.

While I am writing about the Arrochar road and motorcars it is just as well to remember that when I was born motorcars did not exist. Through the years they have been developed largely by trial and error by inventive minds. But when they did come they were poor things compared with the modern car. I remember the first reliability trial that took place in Scotland and there were few places as testing as Whistlefield Hill. The trial took place from Glasgow via Helensburgh and Garelochhead to Arrochar and Loch Lomond and back to Glasgow. I am not sure when it took place but it was somewhere between 1899 and 1901. The day arrived and so did the cars. I posted myself inside the Smiddy wall waiting for the fun and as the cars passed me I had a grandstand view. There were all kinds of cars but the ones that attracted me had steam blowing all over them. I have no doubt that I saw the names of most of the cars and I am certain that the majority were foreign like Panhard and De Dion Bouton etc. Without exception, they all stopped on the steep brae just above the Smiddy. Some of them eventually got away and some just stuck and one thing I noticed was that every car had a sprag attached to it and as they passed the sprag was trailing behind the back axle. The sprag was a heavy steel bar attached to the axle and could be raised or lowered by a cable beside the driver. I walked up the hill to have a look and could see the sprag dug into the ground and sometimes the driver gained a foot or two and then the sprag acted again and of course once over the brow of the hill it became easier to keep going. One thing to remember is that cars had brakes on the rear wheels only for many years after this. Many of the cars could not make it and had to back down hill very slowly and have another try. It was interesting to watch the drivers inspecting their engines and I remember one car had a number of incandescent tubes at the side of the engine and I looked at them fascinated. I think the steam cars were on the whole successful in climbing the hill and burnt out clutches were not uncommon. Some of the cars never got up the hill and had to turn back and others determined to have another try to rush the hill waited in long queues for the chance. It was obviously too severe a test for the cars of the time. The motorcar rally was in the future but this was the beginning of a new era.

6. Blanket washing; Sea Trout.

The population of the village was greatly increased during the summer months by visitors from many parts. These visitors took furnished rooms or houses for periods varying from a month to a week and there were few houses in the district that did not cater for this lucrative business. Many visitors came year after year to the same houses and there were plenty of attractions. There was a golf course, a bowling green, boating, steamer sails and cruises, loch fishing, splendid walks in the hills and by the loch side and plenty of good shops. It was always a pleasure to me to have additional playmates although a sad day when they left for home.

Before the arrival of the visitors the houses were scrubbed and cleaned and much paper hanging and painting took place, done mostly by the occupiers themselves but the operation that interested me most was the annual blanket washing. Different methods were used but the one used by my aunts was the traditional one. A fine breezy day was chosen towards the end of April and early in the morning a fire was lit on an open space below the road. A huge iron three-legged pot was filled up with clean water and set on the stones surrounding the fire, and while the water was heating the cow was milked and let out and then breakfast was taken. After breakfast two wooden tubs were placed in position not far from the fire and blankets were carried down in a big basket. By this time the water would be on the boil and all was ready to make a start with the washing. Slices of hard white bar soap were put into the tubs and then a bucket of boiling water to each and a bucket or two of cold water added to get the right temperature. All the water had to be carried by hand from the old pump above the road, which was still in active use at the time I speak of.

The tubs were of oak and had hand holes cut near the top edges. They were painted green and lasted for many years. Then the blanket was put into the tub, my Aunt Maggie whose feet and legs were bare stepped into the tub and holding up her skirt started to tramp the blanket. Round and round she would go until the soap suds ran over the side, and then the tramping would start again and carry on until that blanket was thoroughly washed. The blankets were then wrung out loosely and placed in a big tin bath and my two aunts, each gripping a handle, carried the blankets across the field to the burn about a hundred yards away. The burn here is about fifteen feet wide and forms a shallow pool with a slow running current. The blankets were tipped into the burn and as there were stepping stones there was no fear of the blankets being washed away into the more rapid current a short distance beyond.

My aunts then returned with the empty tin bath. As the boiler had been re-filled and was near boiling again another pair of blankets were tramped and in turn carried to the burn for rinsing. By this time the first pair had been rinsed in the current and after being wrung out were placed in the tin bath after the second pair had been tipped into the burn. The bath with the rinsed blankets was then carried to a drying line, which stretched for thirty yards at the meadow side, and there hung up to dry in the breeze and the sunshine. The washing continued throughout the day until every available blanket had been washed and although I never counted them at least a dozen pairs of blankets would have been washed. When dry they were folded up and carried up to the houses and those not dry were hung up the following day. Of one thing I am certain there were no two women in the village more exhausted at the end of the day than my two aunts, but I never heard a grumble from either. The big washing event of the year was over and all that remained to be done was to air the feather beds in the sunshine.

My aunts had been doing this since they were young women and doubtless their mother before them. To me, the small boy, it all seemed an episode in the way of life. I suppose another reason was that no one else in the village had a suitable burn handy to follow the traditional method of blanket washing.

Many years later when living in Portugal during the winter months my daughter and I decided to hire a car and drive especially to see the almond blossom. We left Cascais on the 10th of January and arrived in the south of the Algarve a few days later. On the way we had seen the cork forests, Cape St. Vincent and many other places of interest. As we travelled east we were among the orange groves and almond blossom and one day we decided to take the car and go up into the mountains. We had seen mile after mile of almond blossom and oranges ready to pick and now we were leaving this to go higher up. We had followed a valley in which ran a fairly big stream or small river and before long we stopped. On the riverbank there were women of all ages busily washing clothes and on the banks on every bush were garments drying in the sun. I walked down to the riverbank and watched the women at work, but what interested me most were the clothes floating about in the stream, undergoing the same rinsing as my aunts' blankets nearly two thousand miles to the north. I may say that there are few so clean and neat and well dressed as the

Portuguese peasants. Again there is certainly no better climate in Europe than in Portugal or such hard working people.

While on the subject of burns I have previously referred very briefly to the burn[7] which ran through the meadows close to us. It was the favourite walk of my aunts and Sandy the blacksmith. A burn is like the sea, it is a living thing and living things love running water because they depend on it for life itself. It ran between two meadows for about half a mile in a north-easterly direction and southwards with many twists and turns for another half mile until it reached the sea.

On the banks of the burn going towards its source, wild flowers grew in profusion during spring and summer. A lovely bunch of flowers could be picked in a few minutes but we preferred to stroll along picking a flower here and another there. There were flowers of all colours and varieties and one could pick a bunch of different colours and varieties without having more than one of each kind. I have never seen a place like it for wild flowers. Strange to say we had a monopoly of the place as village children rarely or never came to it and it was unknown to them.

Walking up the burn to where its course gradually deepened, more trees appeared on the left bank and you soon came to the place known to me as the Giant's Hole. Here the bank was steep and the stream ran twenty feet below. To me this was a somewhat mysterious place in my early years but I found out later that there had been a wooden bridge crossing to the right bank of the burn, which was somewhat less steep. This was a continuation of the path which ran from Tim na Cross to the village as mentioned previously. But through the years the bridge had disappeared and only a heavy pole with hinged flaps attached remained across the burn to act as a barrier to sheep making their way from one farm to the other. You could clamber down to the burn at an angle and even then it was by no means easy and as a matter of fact it was still used by some people although one was liable to get their feet wet in crossing. There were now plenty of big boulders in the burn and the banks grew steeper and it was no easy task to walk on the bed of the burn. Fully a quarter of a mile further up, the banks were fifty or more feet high and covered with trees but the falls were now within view. When there was spate one could hear the roar of the falls a long way off. The falls were at right angles to the burn which then rose more steeply. It passed under the viaduct carrying the West Highland Railway and proceeded in a northerly direction for about a mile, where it emerged into the open moor. Near here the reservoir had been built which supplied the water

---

7 Often called Macaulay's Burn

to Garelochhead and Whistlefield. A little further up a pipe conveyed water from the burn to the reservoir, which was adequate except in a very dry season. I never inspected the burn much further but there was no doubt it rose on the hills to the east.

I knew every pool and where the best trout were to be found but I soon found out that there were no fish above the falls. When there was more water I would cut a hazel wand and with a piece of brown thin line and a bent pin for a hook I would fish away for hours losing plenty of worms and getting few fish. And then came the day when I had a few pennies to buy a couple of hooks and a proper fishing line. Then I would catch trout mostly about six inches long. I soon learned that even the smallest trout was easily scared and in order to catch them one had to keep well out of sight and use a long rod

One morning in early August I came downstairs to breakfast and saw in the kitchen sink about half a dozen big trout. In my eyes they were monsters but were about a pound each. My father had been up early because there was a spate in the burn and in an hour he had caught all these fine trout. I had no idea that sea trout could be caught and although I was still a small boy I decided that someday I would catch sea trout too.

A year or so later perhaps I was walking home with my father and a workman and we were joined by Jimmy Reid. Jimmy was the husband of Susie Reid of whom I have previously written. He never seemed to work but he was an ardent fisherman and had been in trouble once or twice for poaching sea trout on forbidden waters. As we crossed the road bridge on the way to the village Jimmy looked over the parapet and gave a shout. In an instant we were all looking too, in the pool below there was a shoal of sea trout. There had been a heavy spate a few days earlier and now the waters had subsided and the trout were on their way back to the sea. In less time than it takes to tell we were all below the bridge. I could only stand and stare but the men waded into the pool and drove the fish to the top end where the fish were trapped because of shallow broken water. One by one the fish were taken out and killed until not one remained. They all got a share and I well remember that Jimmy, who no doubt got the biggest share, stuffed his fish into the mouth of a big drain pipe that ran into the burn below the bridge and there they would remain until he returned from the village on his homeward journey.

Many a happy hour I had fishing for sea trout in later years and I always looked forward to the time of the Lammas floods in August. My brother too became a keen fisherman and it was my youngest brother David who yanked out the biggest sea trout I ever saw landed from the burn. With a big sea rod and strong tackle he grassed a three and a half pounds trout and I personally weighed it!

One thing I found out was that the best worms for sea trout were what we called cramblings[8]. They could only be found in the midden among the old dung and were much redder than the common worms and had light coloured rings throughout the length of the body. The trout loved these and if there were trout about we were sure of a catch. Salmon came up to spawn too but I never took much notice of them and left them alone.

---

8   More commonly known as brandlings

7. Natural History

In my day there was a variety of wild animals about and lots of them came out only at night. There were rabbits, mountain hares, weasels and hedgehogs, but I never saw a badger or a polecat. Roe deer were not uncommon but there were no red deer near although I believe they were to be found on the mountains to the east long before my time, and Maol am Fheidh which overlooks the cottage means in the Gaelic *Hill of the Deer*.

When I was young it was the birds that interested me most. I was very small at the time probably just about school age when an apprentice of my father named Duncan Carmichael came out of the joiners shop and showed me two of the biggest eggs I had ever seen. He had been on the hills on Sunday and had found some nests and these eggs he told me were Whaup's or Curlew's eggs. Later on I got to know the bird on both moor and sea shore. Its piercing plaintive cry would echo morning and night and wherever you were you could hear it. Most country boys like to search for bird's eggs and nests, at least they did in my youth.

It is not easy to remember at what age I began to look for bird's nests but anyhow I got to know all the common birds that frequented my home. I suppose the blackbird and the robin were among the first. Then there was the thrush and the tits and hedge sparrow, the wren and the chaffinch. I think the first nest I found with eggs would be the blackbird, which usually built its nest in a hedge, or possibly the mavis (song thrush).

As I grew older I went further in the woods and saw birds I had never seen before and then I soon procured books about birds and their nests and started to make a collection of eggs and at one time I had a collection of over a hundred different kinds. The first bird in the district to build its nest was the mistle thrush and I once found a nest with eggs in late February. In order to obtain a collection I had to range the hills and woods. Usually a younger brother or two accompanied me and we soon became expert tree climbers because many of the rarer birds and even some of the common ones built in difficult places. For instance the long tailed tit usually built in the fork of a tall slender tree. That tit has a wonderful nest with a hole in the side and I could only get two fingers in to try and count the dozen or more eggs. Another tit, in this case the great tit, built its nest in the cavity below the flat roof of a beehive. The cavity was twenty inches square and two inches deep and it filled it completely with moss and placed its nest right in the middle. It must have taken the birds hundreds of journeys to complete the task and underneath the thin layers covering the frames were thousands of bees working away.

The district had a large bird population. Along the burn there were herons, kingfishers, dippers and wagtails of various kinds. Among the brambles and coarse grass there were many more birds, white (or willow) wrens, buntings and warblers etc., and in the conifers, goldcrests. In the trees there were tits, chaffinches, and other finches, and underneath the banks of burns, wrens.

We had many woods large and small and spacious moors with plenty of grouse and blackcock. Hawks were plentiful particularly sparrow hawks and I have seen more than once a sparrow hawk pounce on a bird and carry it away in its talons. On one occasion a hawk pounced on a hedge sparrow but by good luck the bird was at the edge of a thorn bush and the bird managed to get into the centre of the bush in time. I stood watching the hawk making frantic efforts to get at the hedge sparrow and only when I emerged from the shelter in which I was standing did the hawk give up. As I walked to the bush I saw the hedge sparrow cowering in the bush and there it remained for a long time. Once when I was in Mambeg wood we found a sparrow hawk's nest. My brother Bob who was with me climbed the big oak where the nest was. The next time I saw one of the eggs was fifty years later at his home in Canada.

But my favourite birds were the swallows, swifts and house martins that came to us every year in May. They had travelled a long way but in our district they found all they wanted. Plenty of flies and certainly midges were waiting for them, and finally a place to build their nests. Swallows built on a beam in the byre. Swifts also found suitable sites and house martins plastered their nests on the walls of houses below the eaves. The swallows, with cream and rose breasts, and the all black swifts and the martins with white above the rump of the tail, performed their acrobatics in the air from early morning to late evening, one time high in the sky at other times skimming the roadway in search of food. We knew that when they flew high the weather on the following day would be a good and the reverse when they flew low.

On the seashore there were birds of many kinds, gulls including the large black-backed gull to the black headed and common gull. Scarts or cormorants were numerous and the guillemots or "dookers" as we called them, often got washed up on the shore in stormy weather. Ducks and waders were also numerous. I particularly liked the sandpipers. The sea birds seldom or never built nests that I knew about round the loch, but went to more isolated places for that purpose. At that time I was not aware that sea birds travelled long distances to their favourite feeding places or that they had sleeping quarters to which they returned every night.

I used to wonder what "late birds" meant, but as I look out of the window[9] nowadays when the gloaming comes and I see a flock of oyster catchers passing by, some going north and some south, and at intervals a solitary gull, or maybe two or three looking a bit tired perhaps, I know where most of them are going. Some go to the Holy Loch and some to a field between Blairmore here and Ardentinny. Some of them have come from the Gareloch and early next morning will be back there again. The young birds are trained where to go and when the older ones go north or west to breed again the younger birds fend for themselves. Just a few weeks ago I saw a pair of lesser black backed gulls on the shore and nearby a young one half their size. Occasionally they helped it to feed but most of the time it looked after itself. It was at least nine months old and in a month or two it would be left to fend for itself.

There are two sea mammals that were common enough when I was young and these were the porpoise and the seal. The porpoises far outnumbered the seals and they were gregarious. The local term for the porpoise is "the butcher" and every now and then they could be seen with their bent backs above the water just like a bucking horse. They could be heard sucking in or blowing out the air as they disappeared in search of fish. Only once I saw what was obviously a dolphin jumping the length of itself above the water. In the darkness I have often heard the porpoises blowing and my guess is that they feed day and night. The only other thing I need say about the shore at this time is that although there was an abundance of shellfish the natives seldom ate them, and I must confess that I never relished them.

---

9   At Blairmore

Recollections of Garelochhead 100 Years Ago

8. Bendarroch; The Brownes; The Gibson Hall etc.

I have already remarked that my early years at School were uneventful and I suppose that all the time I was making steady progress in reading and writing etc., but there were plenty of things to do and it was the games we played that in retrospect seem to have played the major part in my early years. Of course the small boys played with small boys and the big boys with big boys but age and size plays an important part in who you play with. Football was the game for boys of all ages as I suppose it still is today. We played football morning, noon, and night and literally kicked the boots off our feet. The playgrounds, front and back of the school were tarmacked and the wide road outside the school was just bare earth. An iron railing divided the playgrounds, one half for the girls and the other for the boys. Before we began lessons we had a kick about for a short time, then again at breaks and at the short lunch break and of course after school. We preferred the road outside the school gates as it was longer and broader than the playgrounds and as it was an accommodation road there was nothing to disturb us.

Then all of a sudden, we ceased to play football for a time and some other game became the rage. Marbles which we called "Bools" got all our attention and we played two varieties that I remember. One was called Plunkers and the other Moshie. With the former you held the marble in your right hand and with the middle finger of the left hand you 'plunked' the ball and endeavoured to hit the ball of your opponent as far as possible from a specified mark. In Moshie a hollow was scooped on the ground and you took turn about to get the marble into the hollow. Of course the ground was rough and it was not as easy as it sounds.

When chestnuts were ripe we collected them and tying one to a string we 'bullied' or struck each other's chestnuts until one had disintegrated.

Another game was cat and bat. The cat was a short piece of squared wood and both ends were sharpened to a point on each side of the square. By striking the end with the bat it would whirl away. Figures from one to four were marked on the sides of the cat and the number of hits you had was determined by the number of hits turning up. These were just a few of the games we played but eventually it was football again.

Bendarroch was the big house of the village and was owned by Mr. & Mrs. Browne. They had a family of seven[10], four sons and three daughters, none of whom ever married. The estate extended to some twenty acres and at one time they had a coachman and horses, half a dozen gardeners and almost as many house servants. But circumstances had changed for the worse. The father had taken his eldest son into his business of marine underwriter and during his absence from business the son had underwritten a ship for a very large sum and had failed to spread the risk. The ship was lost with all hands and the father lost all his money.

But the mother had a private income, which enabled them to keep up a semblance of the style of former years but with a much reduced staff. The family had been educated at private schools in England and two of the sons emigrated to Canada and began farming near Regina in Saskatchewan where they did well for a number of years. I well remember Mr. Browne fishing in the burn which ran through part of his estate but he was by then a very old man and died before I was ten years old. I also remember Mrs. Browne, a distinguished looking lady who lived for a year or two longer. One son was killed in the South African War and on the death of his mother the eldest son came home from Canada to look after the estate and the youngest one took his place.

But it was the daughters that I got to know best and particularly Mary the eldest of the family. She had been trained at the South Kensington School of Art and she taught the children of the village everything she had learned there. There was woodcarving, leather work, paperwork, marquetry, basket making and many other things. She and her sister taught the girls to sew and knit, supplementary to what they learned at school. Classes were held in the now disused stables for all who cared to come. They were leading members of the Women's Guild but Margaret the youngest daughter spent most of her time in London after her mother died. They all regularly attended the Parish Church but their main interest in life was the welfare of the village children. When the Boy Scouts were instituted Miss Mary at once formed a troop and became Scoutmaster. When in my middle teens, I became Assistant Scoutmaster. Prior to that I used to go to the big house on Saturday mornings to have lessons on relief carving and she would often hand me books, from their large library, on any subject I was interested in. One of these was on Grinling Gibbons the famous wood-carver of St. Paul's Cathedral. They had a fine curly coated Black Retriever which used to go for the morning papers. There was usually

---

10 The 1881 Census suggests there were four sons and five daughters.

someone with it to carry the mail but I always liked to see it on its morning walk.

There was a very large garden at Bendarroch of about three acres on the other side of the burn and I could often see Robert who came home from ranching in Canada digging and cultivating the garden as it was much too big a job for one gardener and his assistant. They kept a few milking cows and they had the run of a few acres of good grass on both sides of the burn. Eventually the cows were disposed of because they could not get a gardener whose wife could milk the cows and do the laundry work. The Brownes of Bendarroch were one of these families who lived to help other people and the good work they did for our village can never be properly estimated. The children at any rate were lucky to have had them in their midst and I have nothing but fragrant memories of a family that were a truly Christian family.

A village such as ours is a barren place without some sort of meeting place where the people can get together and have entertainments such as concerts and dances. The Parish Church Hall provided a venue for many events but it had limitations in size and there was another called Moirs Hall[11] which also did duty for some events but lacked size and sanitary amenities. When it was announced that Mr. Gibson of Shandon had decided to build a new hall at his own expense the proposal was greeted with enthusiasm. So in the summer of 1897 the Gibson Hall was opened by Mr. Gibson himself in the presence of a large audience. I was not there but got a summary of the proceedings from my elders.

He said at the opening that he had been thinking about the need for a hall for some years and when he had returned from his last visit abroad and having seen Naples and the Indies, the Grecian Isles and the Islands of the Pacific, in his view none could compare with Garelochhead and its glorious hills and mountains and the beautiful Gare Loch after which it was named. He was proud to be a member of the community and to make this gift. I think much of what he said is still true with the reservation that the bureaucrats of Whitehall have done their utmost since the First World War to destroy its beauty with their naval monstrosities and ship breaking yard at Faslane, fouling the sea with oil and sludge, and bringing to the district people whose way of life has done nothing to add to the community spirit which was once the feature of the village. The Hall, built of fine red sandstone, has been enlarged and improved since it was built, with a sum of money left for the purpose by Mr. Gibson. Even now it fulfills a useful purpose. Many a fine ball and concert have I taken part in, long ago.

---

11   Moir's Hall was in Dunivard Road and has been demolished.

Another attraction for the children was the Children's Guild formed by Mr.T.Wilson of Femicarry House who was also superintendent of the Church Sunday School. It met on Tuesdays at seven o'clock and lasted for an hour. We were kept busy making toys and scrap-books and many other things. I remember making woollen balls by passing wool through a hole cut in a round piece of cardboard and eventually producing balls of many colours. Jumping Jacks were great fun and these were painted to show different kinds of people and of course the girls dressed dolls. There was a galvanic battery and whoever could lift a sixpence out of a basin of water while holding one end could keep it. I never succeeded.

One thing stands out vividly in my memory. One evening I think it was the twenty first of January 1901 Mr. Wilson announced that Queen Victoria had died that day and we were all shocked by the news. Young as we were we felt as if one of our family had gone, as the portrait of the old Queen was in every home.

A treat I often had in the summer months was going somewhere for a picnic. On a nice afternoon, escorted by my Aunt Jessie we would set off with a spirit stove, teapot, water and plenty of scones and other things to eat. We varied the rendezvous but quite often we made for Carloch, a conical hill on the west side of the Loch. The route was by Femicarry Burn and up the hill in a south-westerly direction. On the way up as we passed through the gate leading to the hill and close to the burn was a small built-in enclosure. I was always interested in this little enclosure, which was entered by a little iron gate. Built into the wall was a tablet with the inscription **"Here Isabella Campbell was wont to pray"**. I had heard many stories about Isabella Campbell many of which were untrue because no living person had been born when she died. However, I learned eventually that a pamphlet had been written on her life by the Rev. R. Story minister of Rosneath Parish Church who had often visited her during her last illness. She had been born at Femicarry House, which is close to the burn, in 1807 and died in Helensburgh on the 21st November 1827 where she had been taken to receive medical attention. From an early age she developed a strong religious zeal, which increased with the years, mingled with doubts and fears, which were eventually overcome. She developed the early habit of coming out to this spot beside the burn to pray for at least an hour or more at a time. It is recorded that she and the family went regularly to Rosneath Church on Sundays riding in a cart in which they sat on dried brackens. During her short life she inspired many people by her Christian spirit. There was no church at Garelochhead in those days and the nearest were at Rhu and Rosneath, both Parish Churches. Her father Captain

Campbell and brother Dugald predeceased her by about two years. Her sister Mary gained some notoriety after Isabella's death by claiming that she could speak in strange tongues. Eventually she married an Englishman, who had heard about her, and went to live in England. Isabella was buried in Lochgoilhead churchyard her body being taken thence by sailing boat from Helensburgh.

And now to return to our picnic we had a fairly easy climb and just below the peak we put the kettle on and sat down to enjoy the view of the village below and the mountains surrounding us. Not far away was the place where the curlews and lapwings built their nests among the coarse hill grass. On other occasions, we would go in search of crab apples to Rahane Wood or to Glen Mallon Woods. The latter was usually best but much further away. Plenty of rowans and brambles could be found nearer home and even then when possible we would have a supply of lemonade and biscuits with us.

9. School; Mr. Connor; etc.

I was getting along very well at School and when about nine years old was transferred to the big room where the Head Teacher and his Pupil Teacher taught. The Head Teacher's eldest daughter Nan was one of the Pupil Teachers. The big room was divided by a sliding partition. One day I was sitting at a double desk with a girl beside me when I noticed that the skin was peeling from her fingers. I thought no more about it at the time, although I had a feeling that it might be scarlet fever. Some days later my brother John was taken away to Helensburgh Fever Hospital with the fever and then I learned that there was an epidemic in the village. A week or two later I developed a fever and was in bed for two days. No doctor was called though my aunts knew very well that I had a mild attack of Scarlet Fever. A few days later my sister Barbara was taken away. However in my childish mind I had a horror of hospitals and decided that no one was going to take me away and so I kept out of the way of all strangers and would hide below the Joiner's shop when anyone came near, possibly to examine me. But all was well and I was soon back in school and sticking to my lessons.

For the first half hour the sliding doors were pushed back and the Headmaster took over. After a prayer and a reading from the Bible he produced the morning paper and it was always the "Scotsman" and he would read the leader to the classes. Then he would discuss what it meant. Any unusual words were written on the blackboard and a scholar would be asked to suggest their meaning. After this the sliding doors would be closed and the two teachers would proceed with the syllabus for the day including reading, writing, spelling, arithmetic and so on. Each teacher had two classes. Sometimes they took both together but more often separately.

The School opened at 9.30 each morning and closed at 4.00pm in the afternoon and because of this we had only half an hour from 12.30 to 1.00pm for lunch. The reason for this was that a goodly number of scholars had long distances to travel from outposts like Rahane, Portincaple and Shandon. There were short breaks at 11.00 and 3.00 of course.

I among others had to run home for our dinner and run back again and the meal was always on the table waiting for us. And then back to our desks and the next subject. The infant school broke up at 3.00pm and the senior girls then filed in for their sewing lesson. Between 3.00pm and 4.00pm the boys would have bookkeeping, algebra, French, Latin, mensuration, and physical exercises on different days. The only one I missed I think was Latin although I heard most of all lessons while doing something else and I still remember that the Latin word for a table was 'mensa'. He could even teach Greek out of school to certain people. John Connor was the name of our Headmaster. He came from Northern Ireland as a very small boy and was educated wholly in Scotland. He was trained as a teacher at the Free Church Normal College in Glasgow and afterwards spent short periods in schools near Edinburgh. Later he came to Larchfield School in Helensburgh and after four years there he was appointed Headmaster of our school under Row[12] School Board which was established when the Education Act came into force in 1872.

He was of medium height and stocky and had a brown beard and bushy eyebrows. He had pale blue eyes which often showed a humourous glint and possessed the steadfast gaze of one whose life is spent in the observation of human nature. I can see him sitting at his desk examining some papers and now and again glancing sharply round the room at the scholars writing or drawing. He would detect some furtive movement and rising from his desk would saunter up between the desks, the class waiting and watching surreptitiously his every movement. Then suddenly a boy would receive a cuff on the ear and he would say "Put that away". The boy knew full well he had deserved it.

He had a stout cane, which he used on occasions. He gave you one, two or three on the palm of the hand and believe me they were no gentle strokes. But no damage was ever done.

I remember an occasion when I received three for supposed swearing in the playground because another boy had told him that I used the word 'damn'. On another occasion the same boy stuck a pin into me from his seat at my back and I jumped up and looked back at him. I got a couple for that. I was very upset by the incident. King Edward had had an operation that day for appendicitis, which had led to the postponement of the Coronation. When it came to singing the National Anthem I refused to stand and got three of the best for that and probably deserved them. Mr. Connor was no bully but if he had one weakness it was listening to tales told out of school.

Every week we had to write a letter to him. It had to be on his desk first thing on Monday morning. We could write about anything we liked such as home life, a football match, or a walk on the seashore, birds, or a stroll over the moors, or

---

12   Row became Rhu in 1927.

in fact anything we had done or seen. He would collect all the compositions and later return them, as they were written in our exercise books. I think his blue pencil with corrections would be busy on some of them. This went on year after year and I usually wrote my letter on Sunday evenings as I had too many ploys on the Saturdays.

I well remember writing to him about an experience I had. One evening I was taking a quiet walk up the hill to Whistlefield. It was calm and the stars were sparkling in the sky. Shortly after passing Ferndene, the last house before Whistlefield I heard a most peculiar noise. I heard it at frequent intervals. It was a weird sound. Banshees flashed through my mind and I was just a little scared. On my return I again heard the sound which seemed to rise and fall. Anyhow I described the experience in my letter to Mr. Connor and expected some comment from him but there was none. I expect he was too busy looking at the composition and maybe he thought I had let my imagination run riot. I forgot about the whole affair until one day when walking along the banks of the River Ouse near York, I suddenly stopped and looked and listened. It was broad daylight and on my right was a boggy piece of land. Suddenly the whole mystery was solved. It was a snipe "drumming". I watched it fascinated when the noise was repeated again and again as the bird dived down to near ground level and then rose again like a dive bomber. I had often risen snipe on my walks over the moors when a youth but never had I heard one drumming. The common snipe is called the heather bleater in some parts. The bird rises high in the air and then volplanes rapidly to just short of the ground. The sound comes from the vibration of the tail feathers as it dives and is actually a display flight taking place mostly during the breeding season.

Yes, the writing of the Monday morning letter was excellent training in composition and the use of grammar and if there is a better way I have yet to hear of it. Sometimes a special prize was awarded to the pupil who wrote the best essay for which a few subjects were set.

On one occasion the Parish Minister the Rev. A.S. Grant awarded the prize. I wrote on "What makes a man" and got an excellent book prize for winning. The book was called "Men who win". I wish I could see that essay now. Later he came and offered me a post in his father's business in Glasgow. He came more than once but although my mother approved my father would not agree, as he wanted me to follow in his joiner's business.

Mr. Connor was certainly a born teacher of above average intelligence and what was equally important a public figure taking part in all things pertaining to the welfare of the community. He formed the local chess club and occasionally he would go off to the Atheneum Club in Glasgow and have a weekend leaving the school to catch the 4.05 train. He could have moved to larger and more important schools but he preferred to stay in the village where he was esteemed by all. His wife and family of eight were refined people and I got to know them well. He lost his eldest son, John and the youngest, Norman in the First World War and it is doubtful if he was the same man afterwards.

Mr. Connor was one of the old time 'dominies' who, alas, are long extinct but who made Scottish Education known throughout the world. His pupils could never make the excuse that they learned little from him. He taught them to think and again read and write and I think he gave them a foundation, which too few receive today. We need more dedicated teachers. We need more homes where the parents take an interest in the education of their children and who help their children to learn. And we could do with less television, which gives children a false sense of values and wastes precious hours of study and leaves them bleary eyed and tired the following day. What kind of parents are many of the children going to be? All the scientific training in the world is going to be wasted if they are not trained to think and learn to be good citizens first.

Before I close on Mr. Connor I should perhaps mention that our first singing lessons took place in school. I remember the Modulator being unrolled at least once a week and how we practiced the scales. I remember too, many of the Scots songs we learned for the first time, and, at random here are three: "There's nae luck aboot the Hoose", "The Land O' the Leal", and "Is there for honest poverty". One day H.M.Inspectors walked in while we were ready to begin our singing lesson. One went into the Infant room and the other came into the big room. The Inspector at once took us. He got out his tuning fork and, the book being open at the song we were going to sing, immediately he sang DOH, ME, SOH, ME, DOH, and then when we had the keynote he raised his cane and on we went, the boys singing alto and the girls treble. Maybe we sang another song but anyhow he congratulated us seeming very pleased. The Inspectors never spent much time in our school. They knew the calibre of Mr. Connor too well to waste time, but I long remember his singing DOH, ME, SOH, ME, DOH.

All through the winter months and well into the spring we had a singing class in the Free Church Hall, which held two or three dozen people and it was conducted by Mr. Robert Arrol organist and choirmaster of the church. He was a master painter and decorator in Helensburgh seven miles away, and cycled both ways twice a week, on Sundays for the church services and on Wednesdays to take the singing class. It did not matter what the weather was like, he was there. Every season there would be a new cantata to learn and to be performed in the Gibson Hall later in front of a big audience. I can remember one of these vividly. It was called "Britannia". Jean Connor was Britannia and I was Neptune, both of us carrying tridents and wearing appropriate robes and head-dresses. I might have been about thirteen years old at the time and Jean a year older but I was very shy at the time and at the dress rehearsal I almost funked appearing, much to the dismay of Mr. Arrol and his assistants. I had a white beard and did not feel very happy about it. Anyhow Jean opened and I followed with "Old Neptune am I etc..". I am sure the audience must have had a lot of fun and we received the usual congratulations for our effort. I always looked forward to the singing class and learned a lot from them.

Some years earlier Mr. Arrol had formed a flute band for the older boys and young men but it was now defunct. One evening when I was very young I dimly remember the band marching out of the Free Church gate on its way to march through the village. Someone in the band who knew me handed me a triangle and told me to play it and I fell in behind the band doubtless striking the triangle for all I was worth, and feeling very pleased with myself.

In looking back into the distant past it is difficult to understand why only flashes of events are engraved on memory's walls. They are just like snapshots that one takes of a scene of momentary interest but there is the fact that people, men, women, and children are engraved more clearly than most other things.

I can see the people of those days so clearly, as though they were everyday acquaintances. I have only to think of anyone at random and I can see them as they were, not just one or two, but scores of them, young and old, tall and short, man and woman. They are still walking the roads or performing their tasks. They are still alive in that small village. If I were to meet them anywhere in the wide world I would recognise them and call them by their names. But they would not have grown old because immortals never grow old. I would lift my hat to Mr. Ireland, the Free Church minister or to Mr. Grant the Established Church minister I would do the same to Dr. McRitchie and to every woman in the district.

I would pass the time of day with Geordie McKinlay the roadman or Dan McGregor the carriage hirer or Mr. Hart or Mr. Hepburn and hosts of others but none of them would recognise me as wee Willie Hamilton. I have grown old and I am a stranger to them.

There was no inferiority complex about lifting ones cap and no bowing and scraping to beings of a superior caste. Just a mark of esteem and respect to my elders who were bearing unseen burdens I knew nothing about; but who were always pleasant and helpful to everyone. When I was very young I was sometimes sent on an errand by my mother to John McKellar's wife, a fine old Highland woman whose daughter Kate helped my mother with the washing and scrubbing. She would say to me "Ma poor wee crater, here's something for you", and going to a tin she would take out a coin and wrapping it in paper would slip it into my hand. That half penny was wealth untold to me and I would buy some sweets in Mrs. Robertson's shop at the weekend, the only time I could indulge in such luxuries.

10. Food; Fish, Meat and Milk.

This brings me to the food we ate in those far away days. My mother was a fine cook. She came from Peterhead and her father had been a Greenland whaler. He was a magnificent specimen of a man and I remember going to his funeral in Glasgow with my mother. The doctor who had attended him told them that he had never seen a man with such a physique. He had been on at least one voyage of discovery to the Polar seas.

My mother had been brought up on the fare of the fishers of the North with oatmeal, fish and potatoes as the staple diet but she had been a cook in some big houses before her marriage to my father. Porridge was the breakfast dish with plenty of skimmed milk, which could be had for a penny a pint and sweet milk for two pence, sour milk could be bought by the pint or even gallon. Sometimes the younger members of the family would get sops instead of porridge. Sops were made by breaking up a white loaf into small pieces and placing them in a bowl, boiling water was poured over them with a saucer placed on top for a minute or two, then the water was drained off and some sugar sprinkled on top and finally topped up with hot milk. It was a nice change and the whole could be done in a few minutes. Bread was cheap and a two pound loaf could be bought for two pence three farthings and stale loaves for a penny. Before going to bed we would have meal or pease brose made in a similar way to sops, except with the addition of a knob of butter and a pinch of salt instead of sugar.

My mother loved pease brose but I preferred meal brose. For the mid-day meal we had a variety of plain dishes. Soup and milk dishes were frequent. The soup was sometimes made with flank mutton but more often with a bone. Favourites of the family were broth, lentil soup, and leek and potato soup. Creamed rice was a common dish. The rice was boiled for twenty minutes or so and then strained and again brought to the boil after adding a pint of creamy milk, sugar and a lump of butter. It was put into a large pie dish and placed in the oven with cinnamon sprinkled on top. There it baked for an hour or two until it was brown on top. It was a superb dish. I remember discussing with my brother Robert in Canada the foods we liked best when young and he plumped for rice pudding with mealy pudding a close second. We had of course sago, semolina, tapioca and other milk dishes which we relished. Mealy puddings and haggis were often on the menu. Sheep's pluck and lights were minced up with a chopper, an onion or two, oatmeal, suet, salt and pepper and the whole was put into a strong cloth and dropped into a pan of boiling water. What a fine dish to feed a family of nine people and all for a shilling or a little more. The mealy pudding was even cheaper to make and just as filling. We sometimes had stewed sliced sausage and that cost only sixpence a pound. On a Sunday morning we would have some ham or sausages for breakfast and it was the only day when porridge was missing. But they could make sausages in those days and we sometimes had them for tea. Of course, eggs were very cheap and we were never without them.

On Sundays the mid-day meal was invariably a stew of shoulder steak and vegetables with a dessert of stewed fruit. In those days I never knew of tough steak. It cost a shilling a pound, and rump steak two pence more but the shoulder steak had the best flavour. It was carefully trimmed by the butcher and a piece of suet added without extra charge. Occasionally we would have a boiling fowl or a rabbit and maybe at Christmas a pheasant for the Sunday dinner. Nowadays one never knows whether the meat is from an old cow or a bull, but the butchers of long ago had souls and the old cows and bulls were reserved for sausage making in the poorer quarters of the big towns. Joints were unknown to us and these went to richer people or those with small families.

One thing is certain, meat and many other things have lost their flavour and a butcher some years ago told me that there are ten grades of butcher meat. Nowadays it is a question of buying cheap and selling dear.

There was an abundance of good fish. The fishermen at Portincaple would come round most days in the week carrying baskets of fish on their shoulders. Cod, haddock, lythe and saithe were on display as they sat their baskets down and usually there were some hake too. That was our favourite fish and a good one could be had for one and sixpence or if it weighed more than four pounds it might cost two shillings. The hake was boiled whole and served with white sauce and potatoes. With its firm flesh and fine flavour it was a meal for the gods. There was no great demand for it in those days but nowadays it reaches fantastic prices and is seldom seen in fish shops in Scotland as most of it goes to England. Small hake are to be avoided at all costs as they are devoid of flavour. My mother usually cut steaks from the big hake and fried them. They were delicious.

In the summer time when fat herring were being caught in Loch Long the fishermen would come round with their baskets lined with ferns and displaying these herring at two pence each. They would be fried for tea-time and you had a real treat. My mother would often pot herring when they were more plentiful and they would be used to advantage when required. The herring were rolled and the head put through a slit in the tail and were then boiled for ten minutes and then doused in a dish with vinegar.

The baking of scones took place two or three times a week and although I have tried many times to emulate my mother's skill in making light fluffy scones I have never quite succeeded. I think I preferred the scones better on the second day. When I left for the big town and came home at the weekend and saw the big bundle of scones tied in a cloth I knew I was in for a treat despite the fact that in those days bread was bread and more like the bread you can still buy in France. I remember coming home during the Great War and found her making scones with a mixture of bean meal and flour because flour was in short supply and the scones were as good as ever. Occasionally she would make a roly poly and we all liked that with its layers of jam. Boiled in a cloth, put into a long fish pan one did not need much else after two or three thick slices covered with a sweet sauce made from cornflour.

Dumplings of various kinds were made from time to time and my Aunt Jessie was an expert at making the good old Scottish currant dumplings. Once a week especially during winter months she would make one with the raisins, sultanas, currants, spices and all the rest of the ingredients. I always knew when she was going to make one as I frequented the drapers shop regularly and I would often have tea there. She had a special tin for boiling it in and before lunch she would put it on the fire in an appropriate container and by four o'clock when I came out of school it would be ready and until it was finished I would have a slice every day.

There was no electricity or gas in those days and people had to cook on ranges or open fires. In my aunts cottage they cooked on the hob or on a grid. There was a swee, as it is known in Scotland, which was a strong iron bracket that was pivoted on the grate. Pans were hung on the swee with hooks. Of course all pans had to have strong iron handles. The difficulty was to keep a pot simmering and bits of stick or firewood from the woods was one, although coal had to be the only way if there were no sticks. Long practice made the house wives perfect at keeping the fire right in conjunction with the swee and the hob.

11. Sanitation; Housing etc.; Dress of Women and Men

Coal was cheap in those days. My father usually bought a truck of coal of maybe seven or eight tons at a time. The average price was thirteen shillings a ton with a shilling extra per ton for the carting from the station. I used to go into the nearby wood to cut trees and carry them home for cutting into logs. We liked birch best as it was a good burning wood and they kept the parlour fire going until bedtime. I generally spent Saturday mornings cutting logs and putting them into the cellar. Paraffin was used extensively for the lamps and a paraffin cart came round the doors regularly. We usually burned candles in the bedrooms and they were safe and easy to handle.

Toilets and other sanitary fittings were scarce but most houses had water at least in the kitchen sink. Baths had to be taken in a large tin bath and some people never had a bath from one year's end to the other, but there were plumbers about and improvements were being made all the time. The larger houses had more room available for the installation of baths and toilets and some had their own water supply, which had been built before the public supply had been inaugurated. I remember a larger sewer being put in which was laid through the main part of the village.

The rooms of the tenement houses were small and the ceilings low. Built-in beds were common but brass bedsteads were making their appearance especially in the newer houses. There were, in those days, dirty slovenly people who were born lazy if not poor and who were content to live among dirt. But the greater number were industrious people who were proud of their homes however humble they might be. Bare wooden floors were common in the kitchens and bedrooms, and these were scrubbed weekly with warm water, washing soda and soft soap. There would be a rug or two here and there with a strip of carpet beside the bed. The parlour if they had one was often a deserted room only to be seen on Sundays, but our parlour had a fire on at night during the short and cold days and everyone gathered there, reading, writing and playing games. There was never a dull moment in our house in the long dark months.

The windows of the smaller houses had usually small panes of glass about nine inches square and I suppose the reason was that it was cheaper to renew a broken pane, but the trend was to put in bigger panes by removal of the astragals and this made them easier to clean and gave more light. Only in the milder months were the windows opened. The inmates had all the pure air they needed when they were outside and all except the weaklings lived to a ripe old age, many having nothing more serious than toothache or a head cold all their days.

There were weaklings amongst the people of course but that was to some extent a question of heredity coupled with a poor diet when young. Those who led an active life physically and mentally and were happy and contented with their lot were, I believe, the great majority. There is one last point I would make here and that is too many of the houses were damp and the inmates neglected to improve matters. Apart from the absence of damp courses there was the prevalent practice of building on a sloping bank where although built on rock, the water from the higher levels percolated into the foundations and the joiners were kept busy renewing the floors every twenty years or even less. In fact, the same mistake is still being made today and I recently saw two bungalows of first class construction that were built without a thought being given to diverting the surface water away from the houses. The result is that the foundations are often flooded.

Probably as bad is to allow a bank of soil to press against the back of the house high above the ground floor level. It is sad that the officers of housing authorities are only concerned about the damp course and a layer of concrete below the floor, when they inspect the foundations of new houses. Many of the materials used for damp courses are more or less useless after a few years of use and something should be done to raise the standard. There is heavy rainfall in the West of Scotland and in our village the driest months were April, May and June with September reasonably dry and I can well remember many folk having to carry water for weeks at a time because the gravitation supply had failed. But all that is gone and larger pipes and an additional storage reserve has made life easier for the villagers.

The clothing of the villagers varied with their incomes and vocations but in the main it was made of hard-wearing material. The labourers wore moleskins and cardigans in the colder months as they usually discarded their jackets when working. Women wore blouses and long skirts. On Sundays everyone dressed in their best which was very good. Most men and youths wore blue serge suits or a good tweed and starched linen collars were the rule. The women mostly wore costumes made by dressmakers and the bodices fastened at the back went up to the neck usually with a frill of lace at the neck and on the cuffs. They wore hats of various shapes and these were decorated with flowers or bunches of fruit and of course their faces were covered by veils. Some had large spots and some had small spots at intervals according to taste but in my view, they added dignity to the wearers.

Even among the girls of those days there were no knobbly knees to be seen because their parents were in firm control. They wore flouncy multicoloured dresses of becoming shape and appearance and did not need bare legs to enhance their charms.

As I have mentioned earlier, Sunday afternoon was the time for taking a walk. There were plenty of nice places to go but most people preferred to keep to the hard road as one's clothes had to be preserved and had to last a long time.

One favourite walk of my father was to stroll to Mambeg, about a mile away. Duncan McKichan was the Mambeg pier-master. Duncan had a pony and trap, which he used when coming to the village. He always kept a few pigs and got the swill from some of the bigger houses to augment the pig meal. He was a fisherman and, when I got to know him, his brother Peter lived with Duncan and his family.

Duncan McKichan (The King)

Duncan was known to one and all as "The King" and I soon learned that nearly everyone had a nickname. There was "The Provost", "The Deacon", "The Gauger", "The Hake", "Purkie", "Moosie", and "The Gunner". I have a faint recollection that I was once told that at some distant time in the past a sort of brotherhood had been formed and at one meeting it had been agreed that all members be given a nickname and known by that name whenever referred to.

On one occasion we arrived at the pier and found Duncan sitting in his shed at the top of the pier busy filling his pipe. He was cutting slices of a plug of Golden Bar and packing them into his pipe after rubbing them in the palms of his hands. He would light up and puff away. My father never smoked and as a matter of fact few joiners ever did as it was much too dangerous in a joiner's shop but the few that did would often chew tobacco while at work. All around the walls of the shed were photographs of episodes in the South African War taken from the pages of the Illustrated London News. There were pictures of Lord Roberts, Generals Buffer, Hector MacDonald, Baden Powell, Lord Methven etc. I well remember how Duncan and his guests discussed the implications of this battle and that battle, airing their views on what was right and what was wrong and how the war should be won. During the discussion I wandered down to the front of the pier while they talked about politics and fishing and for the first time I saw a porpoise lying dead. I have already described the actions of the porpoises in the water. I asked Duncan about the porpoise on my return to the shed and he told me that it had been caught in one of his hake nets. Of course, Duncan called it a "Bucher". He told me that it was unusual for them to be thus caught because they were heavy and strong and broke the nets as a rule. In any case, they do not as a rule swim deep and pass over the nets. I know that in Loch Long sharks are sometimes caught in Hake nets and they may be twice the size of a porpoise but that is because they become entangled in the folds of the net during slack water. However, it gave me the chance of seeing a porpoise close by with plenty of time to study its beautifully moulded shape.

12. South African War; Gaelic Speakers; Tinkers; Irishmen; Turf Dykes; Miss Grant of Grant.

This was the time of the South African War and I well remember my excitement at the news of the relief of Mafeking and the boys about my own age would sing with gusto the song of the day "*Fight, fight, fight for General Buller. Fight, fight, fight for all his men. And we'll buy a penny gun, and we'll make the Boers run, and we'll never see old Kruger any more.*"

Duncan McKichan was very friendly with the Marquis of Lorne, later the Duke of Argyll, and had his authority to fish for sea trout on the west shore of the Gare Loch during his lifetime. Duncan took advantage of this to "splash" for trout. I never saw him catching trout in this way but I had some experience of this method of fishing with his brother Peter who went to live at Portincaple, but that story will be related later as it occurred when I was older.

We had a sprinkling of Gaelic speakers in the village but it was rare to find anyone speaking it outside their homes. Few of the younger members could speak it as they had been born in the village. But they could mostly understand the language of their parents for obvious reasons.

The crews of the steamers were mostly from Skye and similar places in the West Highlands and when together conversed in Gaelic. I was one day taking a trip to Ardrishaig on MacBraynes steamer and on board was "Erchie" the oldest son of a local family. If I remember rightly he was going to a funeral and of course I kept close to him. We were sitting outside the front cabin when some of the crew stopped near us and began to talk in Gaelic. I noticed that Erchie was listening to them and I could see a faint smile on his face. Now I had known Erchie since I was a child but had never once heard him utter one word of what to me was a strange tongue. The men eventually went off to their tasks and I took a chance and asked him if he had understood their conversation. He at once admitted that he had. It reminds me of a story told by Dr. Norman McLeod. He was of the firm opinion that the Highlanders were the most honest people of all the peoples he had met in his travels abroad. But like guides and ferrymen the world over they thought that strangers on holiday had money to spend. A friend of his who had been a long time in India came home to visit the land of his birth. He had lost his Highland tongue but not the language. He got onto a ferry that would take him home and asked a ferryman what it would cost. This man went off and spoke to the master who could not speak English. Speaking in Gaelic, the master told the ferryman that the fare for the Sassenach would be ten shillings. Returning to the passenger the ferryman said "He says that he cannot do it under twenty shillings, and that's cheap." The offer was apparently accepted, but sometime later the passenger spoke in Gaelic where upon the ferryman rebuked him in the same language. "I am ashamed of you", he said "I am indeed, for I see you are ashamed of your country. Och, to pretend to me that you were an Englishman; you deserve to pay forty shillings - but the fare is only five!"

To me it is strange that Gaelic has several dialects. I was speaking to an oilman a few years ago and he told me that he spoke nothing but Gaelic until he was eleven years old. He had been brought up in a small village south of Oban and one day he went to Campbeltown and confessed that he could not understand half of the Gaelic spoken there and they could no more understand him. This is understandable, more perhaps in my boyhood days than now. A village was a relatively isolated community having its own social life and character and there was little communication with any place five miles away. We might see an occasional traveller or have summer visitors but overall we knew nothing about what was happening there except what we read in the local newspaper.

We had several Irishmen living in the village but with one or two exceptions they were all single men and most of them worked on the railway. They had come when the West Highland railway was being constructed and had stayed on to work on the finished railway way as plate-layers.

They lived in lodgings of the cheapest possible kind with two or three in a room and I believe that they often cooked their own food. I might have been eleven or twelve at the time I write about, and had taken a part time job as a messenger boy and on Friday nights would have the task of taking the Irishmen's groceries to their lodgings which consisted in the main in potatoes, onions, cheese, oatmeal, bread, eggs and bacon. They always ate the onions raw but they seemed to thrive on this plain food, as they were all men of fine physique with no surplus fat about them. These men were certainly a credit to their country and I never saw one under the influence of alcohol. Their wages were only nineteen shillings a week and as far as I know they seldom, or never, returned to their own country. There were no holidays with pay in those days. One of them I knew well. His name was John Leckie and he stood six feet six inches tall and was broad in proportion. He was one of the gentlest men I have ever known and was the ganger of the flying squad on the railway, that is the special gang which moves about from section to section, and his beat lay between Craigendoran and Crianlarich, a matter of forty miles. To hear him speak was a delight to me as I listened to his soft brogue.

About this time I was taking a greater interest in Natural History. There was so much to see around me and to some extent in the soil and the rocks. There were the great boulders that lay about and I used to wonder how they came to be there and by reading I found out that they were the remains of the Ice Age. I used to ponder over this. The rocks always pointed in a north-easterly direction and lay at an angle. But I was interested too in something of more recent origin. The remains of the old turf dykes were common enough and I realised that they had been used some hundred or more years earlier to mark the boundaries of the crofts that had existed on the gentler slopes at the head of the Loch. Over time these dykes had lost some of their original height and broadened at the base. There were obvious gaps where soil had been removed and carted away. In some cases they were covered by trees but in others they were covered by grass or brackens. Digging into them one found a rich brown loam without stones. The soil was excellent for gardens and when my father built his house[13], owing to the position on the side of the road, much soil was required to raise the level of the front garden. He therefore employed a local contractor to cart many loads of the turf dyke soil for levelling up. He then got the idea that he would raise polyanthus plants from seed and as he had a garden frame this was easily done. In due course the seedlings were planted out in this rich soil and the result was a magnificent show of these beautiful flowers and probably that is the reason that it is one of my favourite flowers to this day. One must admire the tenacity of those old-time crofters in cutting and building these turf dykes and then cultivating an acre or two of the enclosed land with a patch of barley or kale or even potatoes.

Wandering tinkers were very common in my youth and a band of them would pass by wheeling an old perambulator containing their belongings. We of course knew where they would be camping and it was not long before the women folk came round the doors offering some of their tins or toasters or some such kitchen ware. But to some extent that was just an excuse for doing a bit of begging. Even if one

---

13   Daisy Bank which is more or less opposite Oakfield Cottages

bought a tin, they never failed to ask if you could let them have a puckle tea, or a puckle sugar, or a wee bit bread, or an old pair of shoon, or an old petticoat. They were most persistent and the women of the house would eventually shut the door in their faces. There was always an old Grannie and she, like the other women, had a dark tartan shawl wrapped around her. Some women carried babies in these shawls and trotting behind were children of all ages in nondescript garments. The men never came to the doors at any time and their job was to erect the primitive tents on their camping site, which was a comparatively easy task. They kept a supply of hazel wands sharpened at each end, and sticking both ends in the ground formed a semicircle with a foot or so between each wand. The canvas was then stretched over, and down to the ground, and after putting in a bedding of brackens or heather their sleeping quarters were complete. A family of five or six could squeeze in and lie side by side. When the supply of tins diminished, the men would fashion new ones or would repair leaking kettles or other tinned goods. They were generally skilled craftsmen inherited no doubt from their fathers and the articles they repaired or manufactured were substantial and sound. If they had a fault it was their weakness for strong drink and of that they could not afford much.

The tinkers, men or women, were never rowdy or insolent. They of course knew that the policeman's eye was on them during the three days they were on any site and I firmly believe that no class of people were more honest than tinkers. They must have been drenched to the skin very often and at other times eaten with midges but you never saw a tinker with a cold or with rheumatism or spots on their faces.

One interesting thing about tinkers is that they always camped beside a burn. We had various burns around the village and one was always known as the Tinker's Burn. Sometimes their site was near the road, and sometimes perhaps a hundred yards from the road, but always secluded and not easily seen by passers-by. It was said that when a tinker's baby was born, the first thing that was done was to immerse it in a pool in the burn and, if it survived that, it passed the tinker's test of fitness. Tinkers always slept in their everyday clothes except their boots and headgear and only changed them when they managed to beg something better. There are not so many tinkers on the roads these days but occasionally one encounters one or two of the kind I used to know. If by chance I have a word or two with them I learn that they are members of one of the clans I knew in my youth. They are true nomads like their progenitors and have been tramping the roads since the dawn of history. Care free and with stout hearts the open road is to them the only life worth living. Most of them take shelter in towns in the winter; some have ponies and carts and some have even dilapidated motor-cars but when the smell of spring is in their nostrils they are ready and waiting for another venture. It is the time for renewing old friendships, for visiting their old camping sites and having again a sleep on a fragrant bed of heather or bracken. These people are not gypsies, who invariably live in caravans, but are a quite separate race. Long may they roam the roads of Scotland.

Saturday evening was the time when many of the village folk did their shopping and about seven o'clock my Aunt Maggie would set off. For some years, until I was ten or eleven years old, I would accompany her. The shops in the village never closed before eleven o'clock and so there was plenty of time to shop and have a talk with friends. At that time she had a nice black cat and in the dark evenings the cat always followed her and seemed to know the time because it was always waiting at the door. When we reached a certain place at the top of Browne's Brae the cat would scramble over the wall and it did not matter how long we stayed in the village the cat would come back over the wall at the same spot when we returned and would follow us all the way home. That cat knew her either by voice or smell and it was far too wise to risk appearing in the village street. Sometimes we would take a walk down the lochside in fine dark evenings before going to bed and the cat would follow us all the way unless someone approached, when it would disappear temporarily.

On our evening walks we would meet plenty of bats and I was always scared that they would strike me in the face as they flew so low at times wavering about and turning in their own length. The common saying "He's as blind as a bat" was well known to me and it was many years later before I learned that the bat has excellent sight. I used to think that there was only one kind of bat but when visiting a gamekeeper in Yorkshire he showed me, in a glass case, twelve different kinds of European Bats. He was looking for the thirteenth.

Sometime in the early nineteen hundreds two people arrived in the village and took up residence at the Anchorage which was a nice house on the Shore Road and had a lovely garden. They were father and daughter, Colonel Grant of Grant and Miss Grant of Grant. The Colonel was a tall upright man with a well-trimmed white moustache and was evidently up in years. Miss Grant was a big woman and had aristocratic features like her father and in a short time was known to everybody. It was not every day that we had real aristocrats in the village and very soon their influence was felt. She took an active part in village life and among the first things she advocated was the installation of street lighting in the village. Maybe there was some opposition but when she suggested that the whole thing could be done free of charge the enthusiasm for the venture was obvious. In those days roads were rough and on dark nights the older people found great difficulty going to and coming from meetings. When one came out of a well-lit hall it was sometimes only possible to find the way home by keeping to the side where there was a wall to touch. And as far as I remember there were no electric torches in those days and it was no uncommon sight to see people carrying storm lanterns.

So, Miss Grant set the ball rolling by organising a series of concerts, which she had advertised as "Grand Concerts". She obtained well-known artists from various centres and between singers and conjurers, illusionists and comics, the concerts were a great success and the Gibson Hall was packed. There may have been private donations that I have no knowledge of, but within a few months enough money had been collected to purchase all the necessary fittings. As far as I remember about two dozen lamp-posts were erected at suitable places throughout the village and its outskirts. The light had to come from paraffin lamps as there was neither gas nor electricity available. Except on stormy nights when some of them blew out, they were a great success and remained until the time many years later when electricity came. A man was appointed to light and extinguish the lamps. The lamps were not lit during the period just before a full moon and for a short period after, and not at all during the summer months.

For some years Miss Grant served the community in every way possible and when she left because of her father's health it was a distinct loss to everyone.

There were in the village however several public-spirited men who worked hard for the welfare of the people. Among these were the Rev. W.E.Ireland, James Hart, and John Connor. A Mutual Improvement Association had been formed by my time and was a very active and thriving institution. It was held in the Parish Church Hall. It was in fact the local parliament. Everyone was welcomed at the meetings. I am not certain when I first attended but it was, I think, shortly before I left school. There were lantern lectures, debates and discussions on every conceivable topic. I remember a discussion inaugurated by Peter McKichan on the Government's decision to put the new Torpedo Range in Loch Long near Arrochar. Peter strongly objected because no fishing was to be allowed in the upper part of Loch Long and of course he maintained that the livelihood of other fishermen and himself was being taken away. While sympathising with the fishermen, the majority however seemed to be of the opinion that the harm done to the fishing would be less than Peter feared, and so it proved.

Questions about water supply and sewage were raised and debated from time to time and the County Council pressed to take action. The Rev. Mr. Ireland was our County Councillor and helped in every possible way to improve the amenities of the village.

Mr. Hart was an Englishman who had come from Manchester as a time-keeper when the West Highland Railway was being built and had married a local girl. He was an able man and took part in all public affairs and was the Parish Councillor for the district. He and his wife had a grocer's shop, but in addition, had interests in poultry keeping. Later, he built a new house at Oxford Place and had a newsagents and sweet shop there. He was a Yorkshireman, born in Leeds but had spent a lot of time in Manchester. He was a staunch Liberal and as I grew older I had many discussions with him in his shop, mainly on politics but often on musical matters. He told me often that Winston Churchill was the man to watch. He had gone to Glasgow to hear him speak on many occasions and advised me to go if I had the chance. He was very friendly with Mr. Brooman White of Ardarroch who was just as staunch a Conservative. That seemed strange to me as a lot of people in the village did not like Mr. Hart just because he was a Liberal.

But I think Mr. Hart was above all a musician and a great lover of Church Music. He was the conductor of the Parish Church choir and a fine bass singer. Every year he would arrange to have a choral service in the Church. I had joined the choir and was a member for a long number of years and often went with Mr. Hart to his home to sing hymns and even some secular songs while he played the organ. I remember at least two Oratorios, which we gave in the Church after intensive practice and to augment our choir we had the assistance of some Helensburgh tenors in which we were rather under strength. I think the first one was Hayden's Creation and the second was Handel's Messiah, omitting some bits of both. Mr. Hart had been brought up with that kind of music and although we were a good choir, we lacked experience in that area. Nevertheless, we had good audiences and were complimented on our efforts.

In those days attendances were always good and, in terms of volume, the singing in the Church was wonderful.

There was usually a Sunday School picnic every year but there was always a Soiree at Christmas. Both Churches had one and always in the Gibson Hall and it was a great event in the children's lives. There was plenty of room in the Gibson

Hall for both children and parents and a large Christmas tree was usually provided by residents such as Willie or Bob McCall who had fine gardens. The Christmas tree was decorated in the usual way and an abundance of presents hung from the branches. Each child was provided with a bag of buns and sweets and a mug of tea. Afterwards there were games and singing and everything was done to make them happy. Finally, Father Christmas presented the prizes and everyone went home happy and contented. I must confess that I often went to both Soirees although I only got a prize at one. I was not the only one who attended both functions.

It was unusual to have a choir trip but one year the idea was mooted and everyone thought it was time we had one. The arrangements were left to Mr. Hart and at small cost we had one of the most enjoyable outings in my experience. A suitable day was chosen and soon after 9.00 a.m. between thirty and forty of us took the train to Tarbet. Waiting for us outside the station was a coach drawn by six horses and believe me my eyes opened wide when I saw them. We all got into the coach and proceeded through Arrochar and round the head of Loch Long and soon were at the mouth of Glen Croe. From there it was a steady climb and I found out that we were bound for what is known as the Little Rest somewhat short of Rest and be Thankful. When we reached the place we all got out. There was a farm-house or shepherd's cottage very near and there we took the food and equipment we had brought with us. I suppose the first thing we did was to make some tea and have something to eat. Then we played games for a time, which were of the kind suitable for both sexes and as we were a choir we doubtless had a spell singing. All around us were mountains especially to the north and south. The scenery was magnificent with the Cobbler to the north and unknown mountains to the south. After a time we had another picnic and were lolling about when someone suggested that it would be a good idea to climb the range of mountains to the South. No one took the idea seriously except Mr. Hart and myself. I am certain that Mr. Hart had never climbed a mountain in his life but had plenty of experience on the lower hills around the Gareloch. And although he was past middle age the two of us set off. We crossed the river and started our climb. On the whole it was not too difficult but our shoes gave us trouble as they were not made for climbing. However, we eventually reached the top of the highest peak. I suppose that we made it under two hours. Down in the valley of the Croe we could see the rest of the choir like black dots and when we looked around there was Ben Lomond so near us to the east that one felt it could almost be touched.

The best view was to the south. We could see the Ailsa Craig maybe fifty miles away with Goatfell and Arran in the near distance. To the west there was Ben Cruachan and, more to the north, part of Ben Nevis and some of the Western Isles. We strolled around for a bit and I found in a hollow what I thought was powdered gold and filled a matchbox with some of it, my intentions being to have it examined by some competent person. But I lost the box somewhere and I later decided that it was fool's gold.

We were not long in descending the mountain and after describing our experiences we had something to eat and drink. At the appointed time the six in hand arrived and we were now on the homeward journey. What I do remember clearly was that we sang all the way to Tarbet station. The song I remember best was "Phil the Fluter's Ball" sung like a real Irishman, which he was not, by Mr. Wilson, a tenor who had been our guest that day, and so ended a day long to be remembered.

When on the subject of church choirs I might say something about the churches. The Established Church of which I was a member was situated near Glencairn Terrace and although it had not much of architectural interest about it, the interior was tastefully decorated and it was commodious and had a fine gallery. The Church was nearly always well filled and one reason for that was that the congregation was very particular about choosing a minister. He had not only to be able to preach well but had to attend to the visitation of his parishioners. The acoustics of the Church were excellent and even partially deaf people could hear and so people would walk miles to the services and feel refreshed afterwards. The services were at 12 noon and 6.00 pm. We had central heating in the Church and it was very pleasant to come into it out of the cold of a winter's day. There was a fine pipe organ, which had been obtained through the Carnegie Fund and Miss Ure was the capable organist most of the time I was there. She taught the piano and I often went to her home to have lessons in solo singing.

In those days there was no radio to give the correct time or even telephone and most people depended on the church bell for the right time and for that reason the bell rang for fifteen minutes in which time people could walk about a mile. There was great competition between the Beadles of the two churches about time. They tried to meet each other before the morning service to decide the correct time. One would provide 'steamer time' and the other would suggest 'station time'. The best estimate was corrected every week day at 1.0 p.m. from the 1.0 p.m. gun on

Edinburgh Castle[14]. They seldom began to ring their bells at the same second but on the whole worked in unison. Both escorted the minister to the pulpit in the same manner standing at the foot of the pulpit after placing the Bible and hymn book on the reading desk of the pulpit and when the minister had taken his seat closed the pulpit door and then entering the vestry closed that door and walked outside to the back of the church. I sometimes attended the Free Church and everything in the service was similar to the other. The Free Church bell was sweeter in tone than the other but both could be heard a long distance off. On weekdays watches were corrected at 7.40 a.m. or 2.40 p.m. the time the steamer rang the bell for the third and last time.

Regarding central heating, the furnaces were lit on Saturday afternoon and oftener during frosty weather. They were most efficient and furthermore very economic and I often wondered why the bigger houses at least neglected to install it. I think the Free Church, although the congregation was smaller, had more prosperous people supporting it than ours. From where I lived I had to walk past the Free Church on the way to the Parish Church and I had to walk past many people on their way to the Free Church. I often found this embarrassing as everybody knew everybody else and I wondered why one set of people went to one church and another set to the other. I continued to wonder why it was that this should be so but the main reason was that if their father had gone to one church it followed that the family members would do the same and it appeared that nobody gave the matter a thought. I think I am right in saying that the narrow religion based on the Old Testament had disappeared or was fast disappearing in our two churches and the Christian religion based on the New Testament had taken its place.

---

14 No explanation of the mechanism involved in conveying that to Garelochhead

Garelochhead Parish Church

United Free Church

13. The Roadman: First hive of bees:

We had poor roads in those days as I have already mentioned and they needed constant attention. Our roadman had a stretch of about four miles to maintain from Balernock near Shandon to the Witches Bridge on the Arrochar road. Geordie McKinlay was his name and of course everybody knew him. He was short in stature with bright red cheeks and a sandy coloured beard. He always had a cheery word for the passer by and rain or shine he could be seen with his wheelbarrow and tools filling in the potholes that abounded everywhere. Small heaps of road metal were placed at convenient places and Geordie would fill his barrow and trundle along putting a little here and there and covering it with soil from the roadside to bind it together when a cart wheel ran over it.

The roads were not tarred in those days but occasionally the big steam-roller would come along and a fresh covering of road metal would be rolled into the worst sections. There was always plenty of mud on the roads too, in addition to the loose stones, and Geordie would draw it to the side of the road and then place it in his barrow and dump it here and there. Where there were gardens near most people were glad to have the road sweepings as there was always plenty of horse manure in it and it was excellent for gardens. In the dryer months there was plenty of dust on the roads and in windy weather it blew about and covered hedges and grass with a grey coating of soil. When the first motorcars began to run on the roads dust became a nuisance and loud complaints were heard on every side and many hard words were uttered as they passed by. As a matter of fact, the cars and their drivers got a share of the stour and if two cars passed each other the drivers were choked or blinded and often had to stop until the air cleared. It must have been a happy day for all concerned when tar macadam began to cover the roads.

After a wet night Geordie had to patrol his beat with barrow and tools, clearing out the roadside ditches and removing leaves from the gratings where the water ran across the roads. With his black oilskin coat and sou'wester covering his head he would work away amid rain and storm and when the day's work was done he would leave his barrow and tools at a convenient place ready to carry on the work next day. On Saturday mornings he had a job to do which he had done for years and that was to clean the village street from the pier to the west side of the Free Church. All papers and loose rubbish were carefully swept up and all made spick and span for the churchgoers on Sunday because that was the day when most people travelled the road. Geordie McKinlay was a quiet, shy man and lived in a house at Rowmore with his wife. Both were much respected and they never missed the morning service at the Parish Church. He has a warm place in my memory.

In passing it may be of interest to know that all the stone used for repairing the roads was quarried locally. Great heaps of hard whinstone were taken to selected sites where it was broken up into small pieces by a man who was known as the Knapper. This is an old Scottish word given to the man who knaps or breaks stones to size suitable for road metal. It is a highly skilled job. This man had a few suitable hammers and with these he broke the lumps of whinstone standing all the time in a bent posture. There was a place on the Knowe, which I have previously referred to, where I saw him at work. It was obviously hard work but when he was finished there was a long regular heap of finished road metal with sloping sides and a flat top. He was paid by piece work and the harder he worked the more money he made but today the Knappers are a thing of the past and giant crushers have taken over to make tar macadam.

One day I met an old lady whom I knew well near her cottage. Chatting to her I learned that her elder son was, in a few days, emigrating to South Africa. He was the manager of a local grocer's shop and much older than myself. I was interested and enquired about the bee-hives he had for I had seen him among his hives at times. She replied by asking me if I would like to have a hive of bees and, if I did, she would let me have one for ten shillings.

I eagerly accepted her offer and told her I would come that evening when all the bees were home. I was still at school and was just fourteen years old but from an early age had dreamed of keeping bees. That evening I had the bees with the hive in my Aunt's garden and placed near a south facing wall in a position that would not interfere with her gardening operations. My aunt was no novice with bees and she produced an old worn out smoker used for subduing the bees and a piece of black net with which to cover my face. I was already feeling important as the bees' new master. I got hold of a guide book within a few days and read it from cover to cover several times and soon got some idea of the management of the bees.

They were a fine lot of bees and I set about making up the section boxes in which the honey was to be stored and I think the placing of these sections in position on the hive was the first time I had made actual contact with the bees. At every opportunity, I would sit at the side of the hive and watch the bees flying out and in and distinguishing the drones from the worker bees. And about nine o'clock on a fine morning in early June it happened! A swarm had issued from the hive, and when I came home I found, sitting on the garden path, an old straw hive and hundreds of bees flying around it. By this time I had been told that the bees had swarmed on to a gooseberry bush and after they had settled my aunt had shaken them into the straw skep and placed it where I found it. She told me that I would need an empty hive and so I took one of the hives, which I had first seen as a toddler many years before in the workshop. After nailing a board on the bottom, assembling some frames and sheets of wax on which the bees could build their combs and with some calico covers cut to size to cover the frames, everything was ready.

In the evening when the bees had ceased flying I placed the empty hive about a yard on one side of the one that had swarmed. I procured a broad board about three feet long and placed it in front of the empty hive and my Aunt placed a white table cover over the board. Then with some trepidation I took hold of the straw skep firmly with two hands and when I lifted it I thought there must be a stone in it. But it was just the weight of the honey filled bees. I walked over to the board and with a quick movement of the arms I dislodged the great mass of bees onto the board. They appeared to be rolling waves, as they endeavoured to find their feet, spreading over the board until the white cloth was almost obliterated. My main aim was to see the queen bee and we both were intent on seeing the queen. In a short time there was an almost imperceptible movement upwards in the direction of the hive entrance. This movement gradually increased and as the first bees entered the hive the bees started to run over the backs of those below and soon they were all in motion and all the time we were both scanning the bees from top to bottom and from side to side, and when about half the bees had entered the hive I saw the queen and she was running too. She, with her long slender, tapered body, the mother of every bee in that swarm, had emerged from under a thick cluster of bees and was intent on reaching her new home. I had seen my first queen, one of the many thousands I was to see in my long life among the bees and I was pleased indeed. I did not feel nervous among the bees on this occasion for I knew by reading that at swarming time the bees fill themselves with honey to build comb and are almost as gentle as flies.

My first season was spent mainly in observing the bees from the outside of the hive but I was learning fast and the first summer I took a lot of fine honey from them. There were plenty of good bee flowers in the district and I used to go about looking for flowers that attracted bees and this way I learned to be more observant than ever and I saw things that in the past escaped my notice. I took an increasing interest in the sky and the clouds and in the wind direction because these things had a bearing on the secretion of nectar in the flowers. I found that honey stored by the bees varied in colour from time to time. For instance, honey from hawthorn was dark brown; honey from sycamore was green, and honey from clover was light golden yellow. Some other honeys like willow herb and thistles was almost white and all had different flavours. However, there was a so-called honey almost as dark as ink found at times towards the end of July just before heather bloomed and a few enquiries revealed that this was honeydew. It is found principally on oak trees and is a secretion, sweet of course, from aphids and is worst during dry weather. The trouble is that it contaminates honey from other sources. I used to hear the bees humming as they flew among the oak trees of which we had more than our share. The only thing the beekeeper can do is to move his bees to the moors where the heather is a greater attraction than honeydew.

When I had a years' experience behind me I began to think that it would be a good thing to move the bees to the heather in August because there was a demand for it at a higher price. Although there was good heather within half a mile of where I kept the bees I considered that if I could take them among the heather I would get better results. I had learned at the blacksmith Sandy Gilmour had often taken some of his to the reservoir when he was younger. So, one day I said to my aunt that I would like to take one hive to the reservoir, around which there was an abundance of heather. Without hesitation, she agreed to give me a hand. I had left school by then and I had a job with one of the local bakers.

The first thing I did was to make a hand barrow to which short legs were fitted and then I prepared my best hive for the journey. The bees had to be on the moors by the tenth of August at the latest and therefore one fine morning when the dawn was breaking about 3.0 am the hive was closed and with the hive securely roped to the barrow we got on the road with me leading and my aunt at the back.

The hill is a steep one, as the drivers of the early motorcars discovered, and although the hive probably weighed no more than fifty pounds it seemed to get heavier as we struggled on. To make it easier I had ropes fixed to the handles, which ran over our shoulders behind our necks and under our arms and these took some of the strain from our arms. My aunt was at the time about sixty-five years old but I was the one who had to stop and have a rest from time to time. After half a mile the hill became easier and when we had gone about a mile we left the hard road and took a diagonal course over the moor. The moor was rough with bogs and heather and stones all the way. Sometimes there was a rough track, which I endeavoured to follow but often the track disappeared in a bog. But I knew where I was going and was thinking of the precious cargo we were carrying. One slip and disaster might have overtaken us. I was young but in no way reckless. My motto was, as I learned later when I joined the Royal Flying Corps "Per Ardua ad Astra" (Through difficulty to the stars). At last our arduous task was at an end and on this bright summer morning we arrived at the reservoir.

We lifted the hive over the iron fence, which enclosed the reservoir and placed it on a level lot facing south. All around were countless acres of heather some of which was already in bloom. I then unfastened the perforated zinc with which the bees were confined. The bees came out with a rush and I ran away as I thought they would be angry but I soon learned at it would have been wiser to stand still. Soon the bees were circling around the hive thus marking their new site or orienting themselves. We sat down among the heather and took a rest for a while after our labours, watching the hive all the time. Within half an hour bees began to return with yellowish grey pollen on their legs which they automatically collected as they probed for the nectar in the flowers. Many a time I have had my boots and trousers covered with the viscous dust of pollen brushed from the heather.

Leaving the barrow behind for the return journey we set off for home with larks singing overhead, the call of the grouse in the heather and the hum of bees in our ears. We got to the main road and I at least was feeling that on that beautiful morning the world was a pleasant place to live in, about half past five we were home again.

For many years my aunt helped me with a hive to the moors, in fact till she was seventy, Why she needed to do so I shall never know but it might have been related to the fact that my two younger brothers emigrated to Canada when they were sixteen years old. There were few like her in these days and I was indeed fortunate in having her.

Every Saturday during late August and early September I would go to the reservoir and take from the bees a dozen or more of fine finished combs of heather honey and replace them with empty boxes I had taken in a basket. During the season, I usually took from the hive perhaps thirty or forty combs of the golden coloured honey, with its snow white cappings of wax. My reason for taking away the honey at weekly intervals was to make the bees work harder, to keep them warmer, and to lighten the hive for the return journey, and principally because I could not keep up with the demand. My aunt had put a notice outside that heather honey combs were for sale. People passing by could not resist the invitation.

One of my best customers was a very old lady who lived until her 100th birthday. She was wealthy and once a week she came for an outing in her carriage with coachman and footman on the box and evidently saw the notice. The footman came in and asked how many combs he could have. He came back with a basket and the order was for a dozen. Time and time again and year after year she came back for more. At four shillings a comb I was feeling rich.

About two years later I had learned quite a lot about how to handle the bees and when one day I met the old lady from whom I had got my first hive she told me that she thought she had one hive that had a lot of honey and wondered if I could spare the time to have a look at it and take off any honey it had. I promised to do so and peeping in I saw that they had filled the top combs at least. So, the first Saturday I was free I took my smoker and an old net with several holes big and small in it and departed to her garden where I did not anticipate any trouble. It was about mid-September and the honey harvest was over and I should have known that the bees would resent any intrusion from me. Lighting the smoker, I put the net over my head. With a screwdriver to lever off the first box I smoked the bees and proceeded to try to lever the box from the frames below. I soon found that to be an impossible task. It was glued fast to the box below with comb and propolis, and the smoker went out. I had roused the bees with my efforts and they were now savage and they were getting through the holes in my net and stinging me on the face and neck. So I retreated to get my smoker working again but also to procure a length of fine wire or string to cut through the wax etc. With everything in order again I approached the hive but of course I ran into a barrage of savage bees but by now I had the smoker working well and my net tucked in around my head and neck. I proceeded to cut the box clear using a sawing motion. Eventually I got the box of honey free and the hive was closed.

I had to get the bees out of the small boxes now and that had to be done some distance away from where the hive stood but that was done without much trouble. I handed the honey into the owner's house and went home with at least a hundred stings on my face and neck. It was all very painful but I had had my lesson and after that no hive of bees ever got the better of me although I got many a sting mainly through carelessness. The following day being Sunday I attended Church as usual, but with a swollen face and an eye partly closed and I remember walking down the road thinking what a fool I had been. Such is the impetuosity of youth.

The years passed and I had joined the Royal Flying Corps and left my bees in the care of my Aunt Maggie, I came back on leave sometime towards the end of 1916 and as usual called on my aunt to have afternoon tea. During this meal, a boiled egg and Abemethy biscuits and honey etc. she went to her kitchen press and, taking out a small package, handed it to me. "That's yours, it's the bee-money." I counted it out and there was sixteen pounds, ten shillings in it. I knew by this time that only one hive of bees was alive and I said. "But surely you did not get all this from one hive." "Yes, I got a hundred and twenty combs from it, and most of them finished complete[15]."

I was flabbergasted, although the summer of 1916 was a good one for bees, I had never approached even near to this figure from one hive. I knew that the handling of this hive must have been a difficult task, in fact a stupendous task, for a woman of between seventy and eighty years. Even now I do not know how she did it. Her method of treating a bee-sting was to take damp soil and rub it over the area of the sting, and it never failed. Naturally I followed her example at least in my young days.

Many years later, after she has passed on in her ninety first year, 1 wrote a book called "The Art of Beekeeping" which is still selling. I know the foundations of my beekeeping knowledge were based on the help given to me by my Aunt Maggie.

One final episode before I leave the bees. One evening in early August a few months after I had been demobilised I was attending a social gathering of the returned soldiers in the village. I had a conversation with Jim Dempster who lived in Glencairn House. He had come to the village from some place in the Borders and had kept bees there. As I was always pleased to talk with anyone who kept bees the conversation naturally turned to bees and about the prospects for the heather. He asked how many hives of bees I had and I told him that my total number was one good stock. He suggested that I should take it to the heather to the best site I knew. I told him that it was a long road to go to the reservoir where I always had success. I had nobody to give me a hand nowadays. At once he offered to give me a hand and if the hive was ready we would go that evening. I did not need much persuasion to fall in with the idea and in a few minutes we were on our way to the cottage where the hive stood. I saw at once that he knew how to act. We stuffed a cloth into the hive entrance and I got hold of two poles to slip between the legs. Everybody around was now asleep as it was near 11 p.m. and getting very dark. So,

---

15   All the cells in the frame filled with honey and capped with wax

we set off on a journey that I first undertook as a boy of fourteen years old. This time it was a different proposition and I began to wonder if we had been wiser to have waited for a few hours and made the journey in daylight. But I had not realised what the journey would be like and experience is the best teacher. We climbed up the steep hill road with me leading and, with occasional rests, reached the point a mile from home where we had to go in a north-easterly direction, but it was now rough moorland all the way. We were both strong and active and up to now the going had been good and we had no difficulty in seeing the hard road beneath our feet. He could see nothing with the hive in front of him but I must have been seeing some light from the northern sky and could tell when the ground sloped, or a bog or some other obstacle made me turn right or left. I had to warn him, as we moved on, of any obstacle in the way. We were carrying a most vulnerable burden and one slip and the hive and the bees it contained would be scattered in all directions. However, the great advantage was that it was dark and in darkness the bees kept quiet and even with an open entrance they cluster and do not fly.

We plodded on very slowly, resting at intervals. How long we took to cross that moor I can only guess but it seemed never ending. As far as I can remember it was well after one o'clock in the morning before we arrived at the site I had chosen. We rested for a short time and then the cloth was pulled out of the hive entrance and our job was ended. And now the job was to get home. I thought the best way was to go straight for Whistlefield railway station and I led the way. Before long I was up to the waist in brackens, and over the knees in heather, crashing into bogs, moving to right or left to avoid rocks and boulders and I had to admit to my companion that I had lost my way. But I knew where the north was, for there was just a glimmer of light in the sky, and so we persevered and at last I knew that the station was near. It had been so dark that only the sky could be seen, but now we could see a part of Loch Long. In a short time we were on the hard road again and set for home. I do not know who took that hive of bees home with me but never again would I go in the middle of the night. In later years I took the bees by wheeled transport to a place above Rahane and although the crops were often good it was never like the old site at the reservoir.

14. Christmas trees; Donald the Shetland Pony.

Now to return to earlier years. Every year at Christmas both Churches gave a Soiree for the children and it was a great event in their lives. Soirees had been held in the Church Halls before my time but when the Gibson Hall was opened in 1897 all such events took place there as there was plenty room for the parents and the children. Each Church had its own event usually the week before Christmas and on one evening it would be the Parish Church Soiree and a few nights later it would be the turn of the Free Church children. The hall would be decorated with holly and evergreens and, at the far end in front of the stage, a large Christmas tree from sixteen to twenty five feet high stood, the gift of the McColls of Altnabuie. The present for each child was hung on the branches of the tree. On entering the hall each child received a bag of buns and with a cup of tea took their places directed by the teachers. When everyone was seated the lamps were dimmed and the candles on the tree were lit. Then the name of each child was called and they were given their present. Father Christmas was of course there to present the prizes. Special prizes were awarded to those who had won them in the Sunday School and there were always additional gifts for the toddlers who were too young to be at Sunday School. Each child got a bag of sweets and an orange with their prizes. Games were played, carols were sung, stories were told, and a most happy evening was spent with everybody happy. I must confess that I often attended both Soirees once as a participant and once as an onlooker, but I had tea buns and sweets at both.

In summer there would be a children's picnic and various places were chosen for the outing, but Glen Fruin was a favourite place. The transport was always farm carts lined with straw and it was surprising how many children could get into a farm cart. Once there they got tea and buns and then played games and afterwards returned tired and happy to their homes.

There were plenty of shops in the village and there was no lack of commodities. The hours of the shop keepers were long and on weekdays were open from 8 am to 8 pm but on Saturdays few shut before 11 pm The three Public Houses (one being at Whistlefield) opened at 8 am and did not close until 10 pm and before my time they opened at 6 am. The two bakers opened earlier at 6.30 am mainly for the sale of morning rolls. There were no half holidays in my early years but eventually they closed on one afternoon in the week except during the summer months. Shops never closed for a lunch hour and I used to wonder how some of the folk had time to make a midday meal. The reason for the long hours of the shopkeepers was that they were mainly family concerns and if assistants were employed they were given lunch and tea breaks. Tradesmen also worked long hours but the working week had been reduced to fifty hours in the summer and forty-five in the winter. Vans from Helensburgh came round selling goods such as crockery, paraffin oil, vegetables and fruit, and the bakers and butchers of the village covered a wide area including Arrochar and Tarbet.

We had at least a dozen good shops. Because of that there was always a demand for message boys. I was a message boy in two bakers, two provisions merchants or grocers, and van boy to Arrochar and sometimes to Portincaple selling rolls, getting up at 6.30 am. I suppose I started this work when I was about 12 years old before and after school hours, and all day on Saturdays. One evening I was told I would have to take some groceries to Glen Mallon about 4 miles away and that I could have Donald the Shetland pony and the small dog cart which I had driven before. So, I had a quick tea and harnessed Donald and though I was far too young to undertake such a journey on a cold winter's night I was young and full of confidence and doubtless considered it a great honour to be entrusted with the job.

I drove quite a lot of horses later, when I went to various places with morning rolls or bread. So, I set off, up and down hill, past Ardarroch and Finnart with its steep hill and on to Glen Mallon. There was only one part I did not like and that was the Witches' Bridge but I had Donald and I liked him. He was a gem, and I would get out and speak to him when climbing the steep hills. Of course, it was dark all the way but I knew the road and the journey was completed without a hitch. I took Donald home or maybe he took me home, bedded him down and gave him a good feed of corn. However, I hated going into the stable at night as it was full of rats. There were rats everywhere, some crawling on the rafters, running up the walls and into holes. Mr. Hart bought a mongoose to try to get rid of them but it would have taken dozens of mongooses to clear the rats. There was another stable next door and a slaughterhouse not far away which I think was the source of the problem. It appeared to be a waste of time trying to exterminate the rats as they came from the ships laid up in the Loch and I was informed that on one occasion a drove of rats came up from the shore and crossed the main road in their search for food.

Donald was eventually put out to grass some years later and I saw him occasionally on the golf course and had a word with him..

15. First Grouse drive; and last one on Ardgoil.

One Saturday morning while working with Mr. Hart I was standing in the shop when a motorcar stopped outside the door and Mr. Brooman White got out and walked in. He asked Mr. Hart of he could spare a lad to go on a grouse drive on the Shandon moors. The two were very friendly and of course Mr. Hart agreed to release me for the day.

The car was an Argyll, which he used at times, and it was filled with guests and guns, and of course the chauffeur. Any cars were indeed rare and I had never been in one and on this occasion I was not in but on the car and I was instructed to stand on the running board and hold on to the side of the car. Off we went, in a short time we came to a sharp corner on the old road near Almanarre. I had to hold on grimly or I would have been thrown off to the side of the road. My first car ride was proving exciting, if nothing else. Fortunately, we had not far to go and turned on to the Glen Fruin road before Belmore, and arrived at the top of the road overlooking Strone Farm in Glen Fruin.

Mr. White shared the shooting with Mr. Graham of Ardencaple Castle near Helensburgh and there would have been a party of seven or eight guns. The remaining "guns" soon arrived and were each allocated a butt, which is an open shelter built of turf. I was given the task of being a "flanker" and was handed a white flag on a stick, three or four feet long. This was a job undertaken by those not fit to walk a lot but the Head Keeper was also there. There were three or four flankers on each flank at the ends of the line of butts and the object was to prevent the grouse passing away too far away from the men with the guns. About a hundred yards separated each flanker. It was an easy job but a most important one.

The men who were to drive the grouse were at least a mile away and were strung out about a hundred yards apart in a straight line with under keepers among them to see that they kept a straight line and waved their flags and shouted now and again. There would be perhaps a dozen men in the line. On a signal the line of beaters began to move and shouting and flag waving alarmed the grouse and they rose from the heather and flew rapidly and low in the direction of the butts. Sometimes they went to ground again but inexorably the beaters drove them on. All the time the flankers had to be on the alert to prevent any grouse passing by the end of the Butts and out of range of the guns. The flankers too had to be out of range of the guns and even a novice like me knew that. As the beaters neared the butts the grouse came in greater numbers and guns were blazing most of the time and birds were falling, but many got past either through incompetent shooting or by approaching the butts at a difficult angle. Grouse fly at a high speed and they always have a better chance of escape if they pass between the butts and not over them.

The men with the guns have a loader who loads the guns. A good loader can do the job in a few seconds. He is very important to a good shot who can kill half a dozen birds in less than half a minute if the grouse are within range, and numerous enough. After the drive the dogs bring in the birds and the gentry in the butts can relax for a time. It is obvious that each shooter must have two guns. Meanwhile the beaters must walk on another mile or more but this time they must keep well out of the area to which the birds have flown. Then they spread out again to face the butts in the opposite direction. The drive begins, and many fresh birds will probably now be included in the drive. An hour or more after the end of the first drive birds begin to come in from the new direction and now the guns are ready to shoot.

On this occasion the guns remained in the one place all day and after the second drive lunch was taken. The keepers and beaters had a bottle of beer and roast beef sandwiches but the gentry had a special lunch brought from the big house but I never got close enough to see. I got sandwiches but no beer as I preferred water to beer. With two drives in the afternoon the day's programme ended and I had been to my first grouse drive.

When I grew up and became stronger I went to several of these drives mostly in the hills facing the Gareloch from Finnart to Rhu. For each drive I was paid 5 shillings and, with gentle slopes, it was reasonably easy. But my last grouse drive is the one I have never forgotten.

In some places there was no road to take the gentry near to the butts and so ponies had to be provided to ride to the butts. It was customary to have two drives to the same butts and then move on to butts two miles away. Many of the gents in my day were physically unfit to walk more than two hundred yards over rough moorland and I expect many are the same today. That is why they preferred driving to butts to shooting over dogs which was a job for real sportsmen who had good legs. In this particular case there were no ponies, or gentle slopes. There was nothing but a boat to ferry us across to the place where the drive was to take place, which was in the

Ardgoil peninsula in Argyll. It is a great rocky mountain range, which extends from Corran Bay in the south to Ardgarten in the north. It is jocularly known as Argyll's Bowling Green, but I doubt if there is a flat place anywhere throughout its length. It rises to over 2000ft above Loch Long. The "Saddle" is probably the best known mountain in the range. We were ferried across Loch Long and landed in a small bay almost opposite Ardarroch House. From there we took a diagonal route to a small Lochan above Corran Bay. A few hundred yards from the shore we passed several ant hills, the like of which I had never seen before or since. I had seen the red or common ant quite often, and occasionally seen small ant hills but the ants in this case had made large conical hills from larch or pine needles. The ants were large and black in colour many times bigger than the common red ant. We came to the Lochan which at the time was stocked with trout. From there we spread out in line with our flags and had to walk to the north. Meanwhile the guns had been ferried across Loch Long to a point opposite Glen Mallon about two miles from Ardarroch and taken up their positions on the steep slope above, where a few butts had been constructed for them. The part we had to drive was above the tree line but just short of the mountain tops and covered with heather. Deep gullies crossed our path and steep banks had to be skirted and I soon realised that this was no ordinary grouse drive.

Slowly we plodded on with no chance of sitting down and taking a breather and I wondered who had arranged the whole thing. The terrain was really savage and there were few grouse to be seen but as we approached the place where the guns were waiting we could hear a few shots. We reached the butts and were soon off again this time to a point short of Ardgarten, at the entrance to Glen Croe which marks the boundary of the peninsula of Ardgoil. We had to keep well up the mountain to avoid driving the grouse over Glen Croe and this walk took a long time as we were putting all our weight on the right leg and the ground sloped at an angle of 60 to 75 degrees. The guns stayed where they were and our job was to drive the birds over them. I do not remember where we had our sandwiches but I was very tired and so were all the beaters.

The numbers of grouse and hares etc. shot are long forgotten. I saw no wild goats or red deer on the drive but I know that red deer were numerous there. I believe that when the drive was on they would run for safety to the Loch Goil side of the range where the terrain is steeper, in fact precipitous. I know people who have shot a lot of deer there before the ground was leased to the Forestry Commission a few years ago. I am certain of one thing. I could have climbed the Cobbler twice with less expenditure of energy as on that day sixty and more years ago. I had to walk three miles home but I had earned five shillings and it was an experience, looking back, that I never regretted.

Just about that time I climbed another mountain, on a Sunday afternoon, with a boy about the same age. It was Maol an Fheidh (The hill of the deer). This mountain faces the cottage where I was born. Every day it's different moods dominated my environment. It is just a little short of 2000 feet and is a comparatively easy climb. In an hour or so we were at the top but to my surprise there was another peak behind it, a mile or so further on, which rose to 2367 feet. It was named Beinn Chaorach. From the top, one had a panoramic view of the Firth of Clyde with Ailsa Craig in the far distance and of the mountains to the west and north, and through Glenn Luss to Loch Lomond to the east. I found on our descent that Fruin Water begins on Beinn Chaorach and passes through Glen Fruin and eventually flows into Loch Lomond.

The hillside was bare, with moss abounding, and as I was a keen naturalist I spotted an unusual plant growing by itself and took a sample. I had no idea what it was. I gave a piece to my companion whose father was the head gardener at Bendarroch. A day or two later the father spoke to me and told me that I had found a rare plant. I am not sure of the name he gave it but I think it was Rat Tailed Moss. Many a time I thought of going up to the mountain to search again for the plant and some others I saw that day. But I never could find a companion to keep me company and the years passed and I never had the pleasure of once more climbing it because instinctively I knew that if I had an accident I would have no one to help me.

Near the Pier at Garelochhead

16. Splashing for trout; and netting herring.

When I was about the age of sixteen to seventeen I was having plenty of adventures both ashore and afloat. Although I was an apprentice joiner with my father, there was little work to do and the trade was quiet. I had plenty of spare time. One day I met Peter McKichan in front of the joiner's shop and we had a talk. He asked me if I would like to go splashing with him that night. I readily agreed and so it was arranged that I should join him at Portincaple at three o'clock in the afternoon.

Peter had left Mambeg, got married and settled down at Portincaple on Loch Long where there were more fish than in the Gareloch, and he was doing well. It was a day in early March at the beginning of the trout season. We set off in his rowing boat with the splash net coiled in the back of the boat. I had no idea of the procedure to be followed but we rowed across Loch Long. It is a loch that is so deep hereabouts that if you tie a lump of lead to a gallon can and lower it to the bottom it comes up as flat as a pancake when you pull it to the surface. We passed the Dog Rock at the entrance to Corran Bay then across the bay where a solitary house stood. In fact there was only room for one house and the owners used it only as a summer house. They were "monarchs of all they surveyed" and the house still stands as it has been kept in good repair by its owner.

We soon passed by into Loch Goil and there on the beach we saw old Walter Allison filling a sack with whelks. These were sent off to Glasgow and made a good profit. Old Walter had three sons, all fishermen. His oldest son could also build boats, but the old man could not do the heavy work of fishing in deep water. It was still daylight and Peter told me that we must hold back for a bit until it was dusk as the fewer people who saw us crossing the better. We were heading for a gravelly shore a short way north of Carrick Castle in front of the farm in which my grandmother was born. By this time I had no doubt that we were on a poaching expedition. Although Peter had the trout rights on the eastern side of Loch Long it was a poor place for sea trout because the water was too deep inshore. Peter was a stocky man and very cheery with dark brown eyes and most friendly in every way. He was a local authority on astronomy, which he never tired talking about.

We reached the place where our operations were to take place and got ready. My task was to row the boat and do what I was told. I had been among boats for years and, although fishing boats are heavier than pleasure boats, I did not think it would be too difficult. Up to that point, Peter had done all the rowing. Peter's task was to work the net. First, I must describe what a splash net is like. The net itself is about a hundred yards long and is composed of meshes of the same size used for herring nets. A rope, called the back rope runs the length of the net and extends for a yard or two at each end. Along the back rope at frequent intervals are attached flat round corks and their purpose is to keep the top of the net on the surface of the water. The second rope is called the sole rope and at frequent intervals small round lead weights are attached which keep the net in an upright position. They must be in contact with the hard bottom for otherwise the fish would escape. The only other things required were two heavy stones tied to the ends of the back rope and these would be in the boat already, in the coiled net.

Peter McKichan and son Jock

Peter swung the first stone on to the shore and then paid out the net at right angles. Meanwhile I pulled the boat out very quietly and when I had the boat out far enough he told me to pull parallel to the shore very, very, quietly or I would alarm any trout. He would tell me to keep right or left a little and then when all the net but a few yards was out, he would tell me to turn at right angles for the shore. I would almost ground the boat and then he would swing the second stone on the shore. My task was now to row back to the first stone and all the time splashing with the oars, or plunking, as Peter termed it, to frighten the fish. Splashing is the name used for this practice which, I may say, is illegal.

Perhaps I should have explained that sea trout keep more to the shore on a flowing tide, to obtain the small marine animals, including worms, whereas in day time they go further out and live largely on small herring, mackerel or other small fry, and it is only in darkness that one can use a splash net. As I splashed and plunked with a wrist action I was rowing the boat backwards and we could hear the fish striking the net in their effort to escape and we knew that we had got a good haul. At last I got to the end where we started and Peter got the heavy stone into the boat and then got the two ropes firmly in his hands and was now facing the stern of the boat. By holding on to the top and bottom rope the net forms a bag and it was my job now to pull the boat as close to the shore as I could without grounding it. There is always the danger that some trout could escape by rushing along the shore edge. The trout are not always in the meshes of the net as they can be too big for this and it is in the later stages of lifting the net when most of them come into the net. As Peter came to a fish he struck it on the head with a short heavy stick. After our first run we went on to a new place and from there the performance was repeated. I suppose we made about half a dozen splashes along to what, even in the darkness, looked like a nice sandy shore and eventually we had enough of it and we prepared to go home. Not a soul had turned up and Peter congratulated me on my work at the oars. It had been hard work, particularly when the net was being pulled in, but I was all the time so interested in the operation that I never noticed how tired I felt. I was so pleased that I had learned a lot about splashing and in the bottom of the boat were dozens of good trout and a lot of red spotted plaice and flounders.

So the return journey began. Peter took the bow oar and I took the back, the sea was a bit rough and the boat was heavier that on the earlier journey. We made for Corran Bay and the wind was freshening. I was tired, but as soon as we passed the bay the old boat seemed to smell home and it became easier to pull. I had not had a bite to eat since lunchtime the day before and was ravenously hungry. It was now past midnight and I did not look forward to the long trek home and bed. However, we beached the boat outside Peter's house, which had been built by my father a year earlier.

His wife was waiting for us. I sat down and saw we were going to have a meal. She set the table and put before each of us a big plate of potato soup with plenty of leeks in it and carrot and turnip. I began to feel better now and the warmth of the small room was a great help after our hard and cold work. Peter kept chatting away about stars and comets and eclipses, then I saw his wife putting the frying pan on the fire, in which she placed two large plaice and I nearly jumped out my chair as they started to flap about in the pan. None of your fishmonger plaice here but live fish which no one except the fishermen ever tastes. Of all the meals I have ever eaten or had in my life time, this was the best. Plenty of bread and butter, with tea to wash it down. I have had meals in 3 star restaurants in France and elsewhere but the memory of that meal lingers still and I was feeling good. The journey over the hill and downhill to my home passed as if in a dream.

When I met Peter in later years he would say to me "Do you remember the night we went splashing at Cuilimuich?" and I would say "That I do, Peter, I will never forget it." His son Jock was not born then, and when I see Jock in his motor fishing boat on Loch Long he is now over sixty years old. His son Peter, the grandson of the Peter I knew so long ago, is helping him to fish too. Such is the way of the world.

Sometime after this my father heard that there were plenty of herring in the Gareloch and as trade was slack he decided to renovate an old boat that was turned over near the joiner's shop and so he renewed many of the ribs and put some fresh planks in the hull. I believe he made a good job and after a few coats of paint the boat looked like new. The boat called the "Lark" was taken down to the shore below the drafthouse[16]. There was a good shore at this point and the boat was moored in the usual way. In the summer we would go out fishing for whiting and would usually get enough to feed the whole family.

The news that herring were in the Loch made my father think that fishing for herring would be more profitable. Therefore, he got hold of an old net that was lying in a lumber room near the shop. It was badly holed and torn so he got two fishermen from Portincaple to come and mend it.

We had no bladders to float the net and I was sent to the painter's shop to collect any spare empty gallon tins to substitute for the bladders. We tied these empty tins to the back rope of the net and in this case the back rope of the net had to be six feet below the surface of the water and corks were useless as the biggest corks would not have enough lifting power to keep the net upright. In this case the bottom rope

16 Thought to be a cottage on the west side of the loch close to the corner where houses first appear between the road and the loch.

did not have to touch the bottom. Some nuts, pieces of iron or even bolts would do to stabilise the net. Herring nets are of different lengths and deeper than splash nets and there are technical terms used which we can dispense with for our purpose. Anyhow we got the net in working order and in our case the net had to be anchored at each end because the Gareloch is shallow, unlike Loch Long where the nets must drift tied to the boat.

We set the net at the head of the Loch and tried to keep out of the way of the steamer, which went around the head of the Loch before berthing at the pier. The net was put out in daylight and there are only two times a day when herring go into the net; at dusk and at dawn. We went out in the "Lark" at night to have a look and to do so we had to raise the net to the surface by lifting the back rope to which the cans were attached. We could see the net below and it looked like a sheet of fire. At least that is what the fishermen call it. In reality it is caused by phosphorescence and the fish, as they approach the net, stop and turn back. It is a fine sight as one looks down to the depths below. On the other hand, when the moon is bright there is little or no fire in the water and that is why the drift net fishermen get their big catches just before and after the full moon. There are other methods used in catching herring, such as trawling and ring netting in which two boats are used, but the quality of the fish is lowered and fire in the water is of no importance. My father knew all about fishing and he had nothing to learn about the craft.

After shooting the net at the head of the loch for a few weeks and netting perhaps a dozen or so herring every day my father decided to go to Faslane Bay where he knew in olden days lots of herring had been caught. We shot the net as usual and the first night's catch was moderate. Then we missed a few days and had another go, and in the evening, decided to have a look at the net. I found that the "Lark" was the easiest boat to row that I had ever been in. She had beautiful lines and we soon got to Faslane Bay. It turned out to be windy and so we raised one end of the net and to our amazement the net was full of fish with a herring in almost every mesh. I think the moon was showing for I saw a pair of ring netters, with their dark sails showing, passing by. These were the days when motor fishing boats were unknown. The next morning we got ready. A strong north westerly was blowing and the boat with the wind at her back was soon at the net. We started to haul in the net and shake the fish out on to the bottom of the boat, and in no time were up to our knees in herring. And then it happened. The old net split right down the middle and lots of fish were lost. All the time the stem of the "Lark" was being battered by the gale and we were shipping water. The weight of fish was the trouble of course. We hauled in what was left of the net and then headed for Mambeg pier as I knew that I could never pull against the gale that was blowing but we knew that a fisherman from Helensburgh was tied up there and we thought he would take our catch. After a stiff row, with my father bailing most of the time we reached Mambeg and when they saw our catch old Duncan and the fishermen's eyes popped. He took five boxes with some left for ourselves but I do not know how much he paid my father. For a day or two I was busy at home gutting and salting herring. I was shown how to do it and for a long time we had salt herring for dinner.

17. First game of Bowls, and how to be a champion; King Edward VII.

From time to time we had visits from itinerant musicians. The Hurdy Gurdy man was a source of delight to all the boys. Carrying his organ, with a monkey perched on top attached to a chain, he would move from house to house, turning the handle of the organ. One tune would follow another and when the sequence was exhausted he would start at the beginning again. When we were tired of following him about we would know all the tunes by heart. One such man, who was an Italian, came frequently and on one occasion he slept below the joiner's shop for the night. When my father found him coming out in the morning he was angry. He thought that the man might have burned down the shop by lighting matches. I do not remember meeting him again.

German bands came too but not so often as the Italians. They were good players and we got value for the coppers they received. Sometimes we would be entertained by a concertina or a tin whistle and usually these people played well. A singer or busker would pass by on his travels. I remember a story about a busker who, on his rounds, stopped in front of a cottage and began to sing. "Oh are ye sleeping Maggie". Unfortunately for him the lady of the house was called Maggie and she came out in a rage thinking it was directed at her. The song was a good old Scots song and the poor man quickly moved on, no doubt wondering what he had done to deserve her abuse.

Brakes[17] filled with people came from Dumbarton usually on Saturdays in the summer. They would be unloaded on the road opposite the cottages and the brakes would turn around and go back to the village where the horses would be stabled and fed. Then at the appointed hour they would come back for the people, who usually walked up to Whistlefield where they spent the day. Occasionally a brake would be hired locally for a day's outing going to Loch Lomond via Helensburgh and Luss stopping at Inverbeg for lunch and then returning by Arrochar. My aunts often told me about a trip they used to take in earlier days. All arrangements were made and early one morning they walked to Portincaple where they took a ferry to Lochgoilhead where they had breakfast. From there they took a large brake drawn

---

17   Large horse drawn carriages

by four horses and proceeded through Hell's Glen to St. Catherines where they took the ferry across to Inveraray, and had lunch. On the return journey the sequence was reversed. They had tea at Lochgoilhead before heading for Portincaple and home. A Red Letter Day in their lives I have often thought.

The men of our village were great enthusiasts in almost everything they did, and probably they excelled more in celebrating national events than in anything else. The Jubilee of Queen Victoria; The Coronation of King Edward; The Relief of Mafeking and other events had to be celebrated by a bonfire. When the bowlers won the County Cup in the seventies my Uncle John, who was a leading light on the Bowling Green got a friend to write a poem about the event. I found the poem in a chest belonging to him when I was a small boy. It became tattered and torn and almost illegible but I wondered if any other Bowling Club had a poem written to celebrate a victory. All I remember of the poem now is the following:

> Let bonfires on the hill tops lowe
> And garlands deck the mead
> The laurel wreath encircles now
> The brow of Garelochhead.
>
> McKinlay fire a feu de joie
> Run all your bunting up
> And shout ashore to man and boy
> We've won the County Cup.
>
> Then join with me, where're ye be
> The single or the wed
> And toast the rinks. The winning three
> The men from Garelochhead.
>
> The Cumbernauld and Renton lads
> Have all had to succumb
> Quiet are the Kirkintilloch squads
> Dumbarton too is dumb.

Proud Helensburgh from her brow
Has had the laurels torn.
Etc. etc.

The celebration I remember best was the one given for the Coronation of King Edward VII. Months before the event (delayed for a time by the King's operation for appendicitis) preparations were made. Near Tim na Cross on the way to Whistlefield a site was chosen a hundred yards from the main road and week by week the pile of logs grew higher and higher and broader and broader. Barrels of tar were integrated into the pile and whenever a cart or wagon was anywhere near a wood or other timber was available it was brought to the site and dumped there. It was all done free and nobody paid a penny for anything. Who organised the whole affair I never knew. I have never seen such a colossal pile of wood gathered to make a bonfire. Of course, I was there, among half the inhabitants of the village, looking on. At the appointed time the bonfire was lit. What a blaze it made. It must have been clearly seen as far as Port Glasgow, and further up the Clyde. There were fireworks from the McCall's home at Dalandhui with varied patterns but all down the Loch there were fires and fireworks galore.

A few months later a special train with King Edward aboard passed by on his way north, and I saw the train, but not the King, although I got reasonably near.

The bowling green was situated in a pleasant position outside the eastern boundary of Bendarroch House. I am not sure when it was constructed but I believe it was in the sixties[18]. It was a few yards under full size and there was room to enlarge it one way only. Most of the men of the village were members but it was always a struggle to keep it in good order and it depended largely on voluntary work. In my time a greenkeeper was employed.

The first thing I remember about it was one day, when I was six years old, seeing a cart passing by the cottage and on it there was a very long tree and walking alongside was my Uncle John. I asked what it was and was informed it was a flagpole for the bowling green. It was a fine larch tree perfectly straight, and my Uncle John would have the task of stripping it of bark, then fitting it up and erecting it on its site. A few coats of white lead paint finished it off. All this would be done free of cost to the bowling club. I often used to look at this fine flag pole with its flag flying, and would recall seeing on the cart many years before. As the years passed I was often on the

---

18   1860s

green watching the players and listening to their wild shouts as some players made a good shot and got rid of an opposing bowl.

I must record how and when I played my first game of bowls. I was fifteen years old at the time and had persuaded my father to take me with him to a game which he was going to play in, for the County Cup. The rink he was to play in was going to Renton. Their opponents were Clydebank. We duly arrived at Renton Green and I prepared to watch the game. The bowls were turned out of their bags and laid on the ground and all was ready.

Two trial ends were allowed to give the players a chance to get the feel of the green, but when our skip tried to hold his bowl it fell out of his hand. He had lost the power to grip it. To my amazement he asked if I would play and after obtaining the consent of the opposing side I took his place as lead in the rink. It was my first game and I had little knowledge of bowling, about bias and weight etc. As I realised later a good lead is important because he must be able to draw as near to the jack as possible. This is to give the other three players the chance to build up a good head as it is called in bowling. I should really have played second as then I would have four bowls to play against. Therefore, my attempts to place my two bowls near the jack were abortive. I felt ashamed of myself, but they consoled me by saying that I would improve as the game went on. But the occasion proved too big for me and I had difficulty in getting a good grip of the bowls as my hands were too small for the full-sized bowls I was using. I do not think the other members played well and the result was that we were beaten by a big score. But the boy of fifteen had been taught a lesson he never forgot, that continual practice is what a good bowler needs.

It would be a year later when I became a member of the bowling green and by then I was an apprentice with my father. Every evening after tea I would be on my way to the bowling green and, weather permitting, would play till dusk. I never hung around waiting for someone to arrive but would get on to a rink and practice by myself. Of course, when members arrived I joined in a game and improved rapidly and developed a special skill in what was known as raking, that is, striking hard to eliminate opposing bowls.

Most bowlers possess bowls of their own and a set of four bowls at that time would cost five pounds, which was quite beyond my reach. However, I found four bowls in a cupboard, two of which had belonged to my deceased Uncle John and two to my father. They were slightly smaller than standard but I could grip them well and had complete mastery of them. There were plenty of competitions during the season and I took part in all of them and soon was winning prizes. After two seasons I was promoted to be a skip, that is the person who is in charge of the rink of four players and in the same year won my first championship.

There was a fine friendly spirit among the members of the bowling club and there is no other game that I know which promotes such good fellowship and real pleasure as the game of bowls. On Saturday afternoon there was usually a game between neighbouring clubs at home and away and a good tea was always provided by the home club.

There was no summer time in those days and it was not uncommon to light a candle to illuminate the end being played when darkness fell. The only trouble at the bowling green was the prevalence of midges. There were some big trees to the west and a deep ditch beyond the boundary of the green, but wherever the swarms of midges came from the difficulty was to keep them out of one's eyes. In time we became partly immune to their bites but the village was notorious for midges.

I think it would be the year after I joined the bowling green, but the year was 1906, when we entered for the Rink, or Fours, Championship of Scotland as it is now called. We had a fine lot of players and they beat every opposing rink in the county and then went on to Queen's Park for the finals. They played brilliantly and got to the final on the last day. When only the last end was left to play, they were six shots up, but then they lost their heads. What happened I do not remember but the opposing side gained seven shots and won. There were plenty of excuses but I don't think such a debacle could happen again between expert players.

It has occurred to me that I can give an account of one incident that took place on the green and the antics of many who were normally sober and sedate. The banter often at times bordered on the ludicrous and was always jocular, and at times sarcastic or cynical. The noise could be heard by the deafest onlooker and the players on one rink would be making such a hullabaloo that the next rink would stand and stare and they themselves would in a few minutes be making as much noise as the others had been making. We had two members Bob Hughes, who was the local postman, and Archie McKellar, who was a jobbing gardener. They would be playing together in the same rink. Bob would be leading and Archie would be second player. They were just ordinary players and I do not think they pretended to be anything else. Archie (we called him Erchie) would play a bowl and, as he did so he would kick back with his right leg and make a short hop. If the bowl came near the jack nothing was said but that was the skips job anyway, but if the bowl was too straight Bob would say

"Erchie, you're as narrow as a hen's face", or if too heavy Bob would say "Had you herring for your tea Erchie?" or he would say "Were you clipping hedges the day?"

When it came to last end Bob would listen to the directions given by the third man and if the skip decided on a strike Bob would turn his back and looking over his shoulder watch what was taking place. But if the skip decided to draw, it had to be a winner to satisfy Bob. Archie hardly ever smiled but it was all done in a good spirit and added spice to the game.

One game I remember when I was a beginner was played on Clydebank Green which was a fine fast green. As I stood on the green I saw in front of me John Brown's shipyard and towering up among the cranes and masts there was a magnificent ship with four red and black funnels. It was the Lusitania the great Cunard liner that was to win the Blue Ribbon of the Atlantic and that was to be sunk off Ireland in the Great War.

At that time periodicals were publishing the account of a prophesy that the end of the world was near and would come to an end in a few months. I must have believed that story for as I looked at the great ship I kept wondering why it was being built. Was it an omen? I never thought about it again afterwards but at the time it was all very real.

I played on the green for a few years longer and among my prizes I still have the Gold Medal I won for my third Championship.

18. The Pier and the Steamers; The Surge; The West Highland Railway.

Boats small and large are always of interest to the small boy, particularly small rowing boats, and we had many of them. They were all congregated to the north side of the pier and their owners did a thriving trade during the summer months letting them to visitors and trippers at sixpence an hour. That may seem a small charge by present day standards but a twelve foot boat could be bought new at ten shillings a foot. Each hirer had a small shed in which were kept all the tackle including fishing lines and anchors.

The oldest hirer was Charles Stewart and he was a very friendly old man who never objected to the boys being present and in the evenings we would get busy pulling the boats into the shore with a rising tide or pushing them out on the ebbing tide, or helping with the gangways. We knew that when the last boats came in he would tell us to take a boat out and have a row, but not to go very far. I was not long in learning how to row and skull the boats and it was great fun. I used to watch the mussels being shelled for bait and at that time there were lots of whiting to be had. Not until dusk would I hasten home, being sure of a scolding for coming home so late. Boats were never let on Sundays under any circumstances and owners of private boats would never think of taking them out.

There was one coal boat still in commission in the late nineties, which took loads of coal to the village. She was named "SURGE" and was owned by old Finlay McKichan whose place of business was at the foot of the brae, leading up to the Parish Church and known locally as McKichan's Brae. Old Finlay was a kindly man although it was said of him that he could skin a louse for its tallow. He had a coal ree[19] and a small grocers shop both of which have long disappeared. But the burn which passed by the end of the cottage, and was also known as McKichan's Burn to the boys of the village, is still there.

The Surge was a sailing smack or gabbart and was a heavy squat boat with a flat rounded bottom, which ensured that she would keep more or less upright when beached. She had one mast, which carried a big lugsail and two jibs. She was tarred

---

19 An enclosure for storing coal.

all over and I believe was about a hundred years old and built throughout with solid oak. She got her loads of coal at Greenock and on arrival was beached on the shore almost in front of the Post Office buildings. For a day or two horses and carts would be at work unloading the cargo when the tide allowed and Finlay's coal ree would be filled up. The Surge had plied around the Clyde lochs since she was built, carrying anything and everything that could be taken in bulk. In those earlier years, the big houses would store twenty to thirty tons of coal and this was the kind of boat needed, for it could be beached almost anywhere that was free of rocks. Coal merchants too made full use of these boats and they were kept busy.

Duncan McVicar was the man in charge. Short and stocky with a trim beard and a seaman's bonnet. He had scars of smallpox on his face, and we knew him as Captain McVicar. I think he lived aboard the Surge and was the only permanent member of the crew as the trade had almost gone. An extra hand or two were employed for the infrequent trips to Greenock. Duncan was a "character", a kenspeckle figure in the village and after old Finlay died, the Surge and Duncan passed away and soon became only a memory.

The steam puffers had long before taken over the trade of the gabbarts, and with the arrival of the railway their trade also was at an end, confined to places where there was no railway. The pier was near to the head of the loch and there was just sufficient room for the steamers to turn around without reversing in approaching or leaving the pier. The pier was comparatively new, built in 1845, because the original pier was situated at Rowmore where the water was deep and a rocky promontory made a short pier adequate. Rowmore was at least half a mile to the south of the village and the pier there caused much inconvenience to those who lived to the north and west of the loch. The old pier was built about 1820 shortly after the first steamers began to operate on the loch, and continued to be used for forty years, when it was dismantled. In 1835 the first Sunday steamers were advertised to run from Glasgow to Clyde resorts, and Garelochhead and Kilmun were chosen to try out the new service. The first excursion to our village was on the last Sunday in August and strong protests were made to Sir James Colquhoun to prevent the desecration of the Sunday by prohibiting the landing of passengers at his pier.

Sir James readily agreed to do all he could to prevent the steamer landing passengers. He gave orders for all gamekeepers on his estates to attend on the Sunday and in addition obtained several policemen from the authorities. The steamer, named "The Emperor" arrived shortly after 1.00 p.m. and drew in to tie up but the Captain was told that he could not do so. He refused to leave and after a few failures a rope was got on to the pier and the boat was tied up. Sir James Colquhoun was Lord Lieutenant of the County and he had issued strict instructions that nothing must be done to endanger life and that law must be observed. It was soon apparent that on the steamer there was a band of thugs armed with bats and bottles, and intoxicated besides, and so the defending force was ordered to withdraw as bloodshed was sure to take place if they tried to repel the hoodlums. However, Sir James applied for interdict and in a few weeks this was granted and except for another visit by the Emperor a fortnight later to the old pier at Rowmore no Sunday steamer ever again landed passengers at Garelochhead.

Lady Clare at Garelochhead Pier

Captain McKinlay was in charge of the first paddle steamer that I remember called "Lady Clare". He had been Captain of "The Gareloch" which I do not remember. He had been a deep-sea man in sailing ships before joining the Gare Loch Steamers and from what I was told he had been on these sailing ships for thirty years or more. He was a tall upright man and resided in a house off the Station Road. I heard many stories but most have been forgotten except one. He could sail his steamer in the thickest fog and could take all the piers to Craigendoran with the aid of the compass, chronometer and patent log, which was something no Captain ever attempted even now. He was due to retire when I knew him first and the Lady Clare was taken over by Captain Carmichael who had been pilot and mate with Captain McKinlay. Some years later the "Lucy Ashton" became the Gareloch steamer under Captain MacDonald and continued the run for a long time.

The North British company had at least seven steamers in commission during the summer months and were usually based at Craigendoran pier. In my opinion the North British steamers were the most picturesque on the Clyde. The hulls were painted black with gold lines and their funnels from the deck being red, white and black, that is red for about three quarters then a broad white band and black at the top. The North British steamers were all well known to me, they were the Waverley, Red Gauntlet, Kenilworth, Talisman, Lady Clare, Dandi Dinmont, Lady Rowena, Lucy Ashton, Diana Vernon and later Marmion.

The Gare Loch run was the most remunerative to the Company. The steamers always lay at Garelochhead overnight and most of the crew resided in the village. The timetable never varied from one year to another. The steamer left promptly at 7.20 am every morning, except Sunday, and arrived at Craigendoran at 8.15 am. Then she normally went to Greenock and returned to Garelochhead at 9.40 am leaving immediately on the same run and returning at 12.30 pm. She lay at the pier until 2.40 pm when she sailed again, returning at 5.45 pm on the last run of the day except in the summer when there was an extra evening run finishing at 7.45 pm. On Saturdays only, instead of returning to Garelochhead at 12.30 pm, she went directly to Craigendoran. Thereafter she called at all piers and returned to Garelochhead at 2.40 pm.

The piers on the loch were Mambeg, Rahane (ferry boat), Shandon, Barremman, Rosneath, Row (now Rhu), and Helensburgh.

There were always plenty of passengers going both ways and a return to Helensburgh was only one shilling and to Greenock one and sixpence. Occasionally I would go to Greenock, usually accompanied by an adult, to obtain stores. Among these would be a gallon of syrup and a gallon of black treacle from the Glebe Refinery there. We used the treacle for the porridge and scones and the syrup for general sweetening, and spreading on bread as a change from jam.

Goods of all kinds were carried by the steamer and the sacks of sugar and meal weighing at that time two hundred and eighty pounds had to be carried by the deck hands into the shed on the pier. It was a tough job loading and unloading and was a job for a strong man. The deck hands and stokers if unmarried lived in the steamer and had to cook their own meals. Their diet was mainly potatoes, herring, porridge and mutton, and broth, but as they were invariable West Highlanders and fishermen, they thrived on it.

When I went to Helensburgh or Greenock I used to spend a lot of time peering into the engine room, watching the huge cranks, and all the machinery and complicated apparatus which turned the paddle wheels. The greaser or engineer would put some oil here and there and the flashing brass dials were mainly a mystery to me. I often wished the engineer would ask me to come in, and explain everything to me, but he never did.

My favourite steamer of those days was the Waverley. She was built in 1899. Once a week in the summer, she would be billed to go on an excursion to Ailsa Craig, around Arran, Loch Long, Loch Goil and Arrochar, Ayr etc. On the bill it would always state 'Weather Permitting" and there was always a fear in my mind that it would rain or the wind would make the sea rough and stormy. I would be told in the evening that we might go for a sail in the Waverley the following day and so early the next morning I would be up with the lark and have my breakfast before 8.00 a.m. I would hurry down to the Knowe and listen if I could hear the paddles of the Waverley coming through the narrows at the foot of the Loch. As soon as I heard them I would run back home and shout "She's coming" and we would set off for the pier. Not until I set foot on the steamer did the heavy feeling on my mind, that she might not go after all, would vanish. Once aboard this fine steamer the world was mine and every part would be explored up and down.

When we arrived at Craigendoran a big party would come on board including the band, who would play on and off all day. The tune I liked best was 'Astuariana' with the background of the paddles blending with the music. I had the advantage that I could follow the band around the ship and heard the tunes played again and again. A call would be made at Dunoon and Rothesay and other places and the steamer was packed with people. The Waverley was as fast as any and would take on the King Edward or Queen Alexandra in a race to Rothesay and the other steamers nearly always gave way as she could do twenty-one knots if pushed.

All the excursions were full of interest but I liked the ones to Ailsa Craig because when we got there the steamer would slow down and rockets would be fired. The rocks would be covered by sea birds of all kinds and as the rockets were fired the sea birds rose in their thousands. It was a wonderful sight. Then as the Waverley turned around on her homeward journey the birds returned to their nests.

In the evening when we entered the Gare Loch with the beautiful gentle slopes around the Loch and the eternal mountains in the background I thought there must be no fairer sight than this. I took it all for granted I suppose not thinking then as I have often done since that I was lucky to be born and live in that place and at that time, another golden day had come and gone and tired but happy I was home again.

Garelochhead Hotel and Pier

There were numerous sailings to the village and I got to know the Glasgow and South Western steamers well. These excursions usually came from Ayr and Ardrossan and arrived in the early afternoon. There was the Juno, Jupiter, Mercury and Mars. I liked the Juno best as she was big and sat high in the water but they were all handsome boats with their light grey hulls and red funnels. The passengers got an hour or two ashore to take a walk around the village. It was a sad day for the Clyde when the North British, and Glasgow and South Western, steamers disappeared with the amalgamation of the railways, leaving only the dingy looking fleet of the Caledonian Company. I may say in passing that I sailed to Ayr on North British steamers to see the land of Burns.

Then we had the "Black" boats of the Buchannan fleet, which came on Saturday afternoons from Bridge Wharf in Glasgow. They must have been old boats with new names and they were well patronised. As they approached the pier, it was not uncommon, and alarming at times, to see the passengers crowding to one side and the steamer heeling over at a steep angle. Nor was it an uncommon sight to see two steamers lying side by side at the pier with a third one waiting outside for the passengers to land. In later years I made the trip from Glasgow in a "Black Boat" just to see what it was like and I thoroughly enjoyed it, all for one and sixpence return.

From time to time the local steamer was relieved by another one of the fleet and this time it was the "Red Gauntlet", this fine-looking steamer was taking the afternoon run and as usual was circling around the head of the Loch to take the pier from the north. Something happened either to the steering or the engines and she crashed into the end of the pier. The people aboard were thrown on the deck and those on the pier ran for their lives. The bow of the steamer was badly buckled and the pier ever afterwards was two feet off perpendicular. Nothing short of drawing the piles would have put it back in position. The Red Gauntlet had a fault, it was generally believed, in the steering, and sometime afterwards she struck a rock off Arran and was sunk.

The Lucy Ashton had one or two accidents and one I remember was that one dark winter night during a snowstorm she struck a laid-up ship near Rosneath. No lives were lost, but some were seriously injured and the steamer was badly damaged.

The Loch was for long periods filled with idle ships and they would disappear one by one and then come back as trade slackened. They were mostly tramp steamers but there were some sailing ships among them. I remember the "Helen A Read", an American sailing ship that remained off Mambeg for some years, and, in later years ships of the Loch Line.

One of these ships, I think the "Loch Ness", was moored near the head of the Loch for a year or two and one day a local man was asked to find some youths or men to man the capstan bars as a tug was on its way to take the ship away. I was among the youths chosen and was told that the job would take about half a day. I was about sixteen years old at the time and was quite strong and sturdy and in any case the ganger had to get people he knew. The ganger rowed us out in a boat he owned. We had no coats or other warm clothing thinking as it was an easy job we would soon be home again. We got on board and started on the capstan bars walking round and round but before long we got to a point where we could not move the chain.

It was then discovered that the swivel to which the chains were attached had rusted and the chains were twisted round each other. The tug could do nothing to help because one tug could not turn the ship around by itself because of the tides and the number of twists in the chain. They telegraphed to Greenock for another tug and expecting it to come in the evening we were all kept on the ship waiting for the tug to arrive. We had brought nothing to eat and as darkness fell I was not feeling too happy except our pay was still mounting all the time. However, the second tug did not arrive till morning. We had no place to sleep as the ship was completely bare and the only place we could bed down was in the forecastle in hard bunks with not even a bit of sailcloth to cover us.

The hawse holes were open with the chains going through them. As the ship faced the wind, the breeze came blowing over us and we were stiff with cold. We got a drink of coffee in the ship's galley before we lay down but that was all. As the night wore on we got up from the wooden bunks and tried to find a sheltered place, but it was hopeless. We got another drink of coffee and someone found a tin of ships biscuits.

That night I pictured myself as an apprentice seaman on the way to Australia and then vowed I would never be a seaman. As I looked up and saw the great masts and spars against the sky I thought to myself that they were brave men and boys who sailed in these ships, and even they had a bit more comfort than I had. It was a great relief when the second tug arrived and by then we were a sorry looking lot. The tugs started, one pulling and the other pushing and for the first time we got into the galley, which all night had been crowded by tug men. There was a fire in the galley and coffee was doled out by the ship's watchman. After a few hours the tugs pulling and pushing got the twists out of the chains and then we set to work again walking round and round. We sang no sea shanties but the anchors came up and we had done our job. We rowed wearily to the shore looking at a fine ship sailing between two tugs, and what did I think? I do not know. What I did know was that I was tired and hungry, but not cold anymore. We were paid in the local pub and I do not remember how much, perhaps a shilling an hour perhaps more or perhaps less. It had been an experience I would not like to have missed, but it was not one I ever wanted to repeat.

At home a good meal was waiting for me and my mother had been much worried at my long absence. I had been nearly thirty hours away without sleep or food and I was soon in bed and sleeping the sleep that only healthy youth enjoys, to awaken next morning to another day of adventure. In this period in my life as I awoke I would look out of the window and say to myself "The sun, it is shining for me. A job to do, something new to see."

The West Highland railway was opened in 1894 and it was and is one of the great railway engineering achievements of the age. We had plenty of steamers and were relatively well off in communications. We could go to Helensburgh and there get a train to Glasgow and beyond, but the new railway gave us a direct route and it opened up the West Highlands to us. Glasgow was little more than an hour's journey and Glasgow people could come to the village cheaply and easily. Garelochhead is, or was, a particularly suitable place for a day's outing. Plenty of fields, gentle hills, shell fish galore, rowing boats and three halls for parties. Soon after the railway was opened excursions came, mainly from Glasgow, sometimes three train loads one after another and hardly a Saturday passed but there was a party. Usually they made their meeting place in the field, used by the football club, in front of our house. Men would arrive early and set up the boilers for making the tea and sometimes a big tent or two would be erected.

The first part of the excursion would arrive at the station in mid-morning headed by a brass band. The sound of the band would make every boy in the village converge on the Station Road and you couldn't see my heels for stour as my father used to say (and there was plenty of stour on the roads in those days). How we enjoyed those bands with their banners flying, playing a march, with the big drum beating the time. I was thrilled with it all and as we followed the band to the field. We hoped it would go on playing all day long. The main village street would be packed with people and all around the shore there would be crowds, some paddling, some gathering mussels and whelks and the loch would be dotted with rowing boats. In the late evening they would wend their way to the station with the band still playing and into the trains and home, carrying the toys and souvenirs bought in the village shops.

We also had excursions to Oban, Fort William, Edinburgh and Stirling. For about four shillings one could have a long day's outing to Fort William. If it happened to be a fine windless day the scenery of the West Highlands, with its quiet lochs sleeping in the sunshine among the great mountains as seen from the carriage window, was never to be forgotten.

I remember going to Edinburgh with my father and brother John to see a pantomime. We stayed overnight, but all I can remember is that a great round balloon like object or lantern filled the amphitheatre with all the colours slowly turning around. The pantomime was probably Dick Whittington.

The railway station was above the village and there was a good road to it, which sloped gently from the top of the village to the platform and there was a stair, which could be climbed to shorten the bend near the top of the road. There was plenty of room and the sidings were filled with wagons of coal and bricks, etc. There was also a turntable for turning engines and engines could be filled with water at the end of the two platforms. In fact, the station was designed to take supplies including bricks, coal and other building materials to a large area.

More and more people came for weekends. We could get a return to Glasgow for one shilling and nine pence on Fridays and two shillings and sixpence on other days. On Mondays the platforms were crowded with people going back to work after their weekend, and dozens of commuters travelled daily to Glasgow. Excursions large and small came to the village throughout the season. Yes, the railway opened up the West Highlands and just one last word. You could leave London about 7.00 pm in the evening and at 7.45 am every morning I could hear the express train passing over the viaduct a mile from my home on its way to Fort William. Year after year it passed and during the shooting season there would be two trains filled with sleepers on their way north.

Garelochhead viewed from the Station

19. Miss MacDonald of Belmore: Faslane burial ground.

About a mile and a half from the Post Office on the Helensburgh Road stood Belmore House. Miss MacDonald was the owner as her mother had died when I was very young.

She was a very wealthy woman and was a benefactor of all good causes and a member of the Free Church. She kept a large staff including a butler, footman, coachman and groom, and a few gardeners as well as the indoor staff. She also owned a large steam yacht, which was permanently anchored in front of the house. In charge of the yacht was Captain McKinven who lived in the village. During the winter the Captain would board the yacht to see all was well, but in the spring the crew would arrive, the deck hands from the Hebrides and the steward, cook and engineer from Glasgow.

Miss MacDonald would have a sail on the yacht about once a week, going down the Clyde to Rothesay or up to Arrochar or even through the Kyles of Bute and around Bute. As the years passed she used the yacht only on rare occasions, but all the time the yacht would be in commission. In the summer evenings the crew would have a walk to the village dressed in their blue uniforms with brass buttons and with white shirts, and the finest of brown shoes. Every year they had a new uniform and new shoes. They were fine upstanding men and of course they came back year after year. The yacht was named "Elfrida" and was painted black and gold and it certainly was a sinecure for Captain McKinven and his crew.

The Captain also acted as a general factotum and could be seen walking daily to receive his orders from Miss MacDonald. I remember once calling at the front door to solicit a subscription for something and being asked into an ante room by the footman. I remember having the greatest difficulty in walking on the very highly polished floor, meanwhile glimpsing into the hall which revealed to me the magnificence of the place.

Miss MacDonald had woods behind the house and she got the idea, so I'm told, that dead leaves were unhealthy and gave orders to her head gardener that all the leaves should be raked up and made into leaf mould or burnt. So, the head gardener collected a team of men including youths to rake up the leaves and clean the woods.

Year by year the work started in late January and lasted for a month. I remember that one year I got into the team and every morning we tramped down to Belmore and came home every evening at six o'clock.

Folks said she was eccentric in having all the leaves cleared away and that it was a waste of time, but I believe it was Miss MacDonald's way of giving much needed work to the men of the village who were out of work at this time of year. I well remember raking the leaves and wheeling them away with the primroses in full bloom at the time. On another occasion on our homeward journey we were passing the pier when one of the party called Sandy Grey (a painter and decorator) said "You can read a sign at 6.00 pm on the 12th of February". And I still remember what he said so long ago, and when the days lengthen I know that on the 12th of February it is dusk at 6.00 pm.

A short distance to the north of Belmore House was Faslane. A quiet place with only two houses of any size and a farm cottage, but it is probably the oldest and longest populated place on the Gare Loch. At one time it was one of the seats of the Earls of Lennox. Vestiges of the old church and churchyard remain, but no signs of the castle and the legend is that Robert the Bruce worshipped here in 1314. It is now enclosed in the new cemetery, which was made in 1903. As far as I know no one had been interred in the old burial ground for a hundred years and more, for Rhu churchyard had long been the burial ground of the Parish, until closed at the end of the century. I used to stroll through Faslane occasionally but there was little to attract one to visit it often, but most of the folks I knew in my youth lie there.

It is sad to reflect that the nuclear submarine base has swallowed the whole place up, leaving only the new cemetery, with its surrounding stone walls as an oasis in a desert. Perhaps one day and it cannot be too soon, the whole will return to the quiet beauty of the Faslane I knew in my youth.

20. The Volunteers and Territorials; Rifle shooting; and the first camp of the 9th Argyll

In the office attached to the joiner's shop, my father kept his books and made out his accounts. There were lots of odds and ends and when I could get hold of the key of the office I used to have a look round. I soon discovered a lot of old rifles and guns hidden in a corner. They were all muzzle loaders with ramrods attached and some of them were beautiful specimens of the gunsmith's art, but somehow I was not interested in antiques. I learned that they had belonged to my late Uncle Bob, who was a first class shot and had been a member of the local Volunteers. There were moulds for making the lead bullets and old percussion caps and powder flasks. I wished I knew how to load and fire them and perhaps it was as well that I did not.

However, these guns gave me the urge to learn how to shoot and one day I was informed that there was a proposal to form a section of the Volunteers in the village and that a meeting was to take place. I was probably about seventeen at the time and, needless to say, I enrolled in the 1st. Dunbartonshire Volunteers attached to A. Coy. in Helensburgh. Once a week the staff Sgt. Major came up from Helensburgh and we were put through our paces and issued with a rifle. We were measured for our uniforms, which consisted of tartan trews and a beautiful red jacket and greatcoat.

On good evenings in the summer we drilled in Moirs Park, otherwise in the hall. The section was issued with two or three .22 converted rifles and a large iron plate was fixed in the hall and on this the targets were placed. I soon found that I could shoot well and in due course won a silver medal with gold centre. There was no outdoor rifle range in the village but on most Saturday afternoons during the summer we would go to Helensburgh and shoot there. I became a regular attender. There were some fine shots in the Company and they gave tuition to novices like me. Among these was T. Kerr who had won at Bisley on many occasions. We could buy our ammunition on the range and there was no rationing. I don't remember how I felt when I fired my first round but it was quite an experience from using a small-bore rifle. I had to have a sliding sight to allow for wind and soon learned the tricks of the trade. I enjoyed every minute I was on the range but strange to say I was sometimes the only one from the village.

Every year there was a Battalion shoot at Auchendarroch range in the Vale of Leven when numerous trophies were to be won, and as I was one of the best shots in our section I decided to go. The message came rather late with the necessary warrant and I had no idea that there were certain regulations to be observed. With my rifle, I duly arrived at the range and was surprised to find that everyone was in uniform. Hitherto I had attended all local shoots in civilian clothes and was now surprised to find that I would not be allowed to shoot unless in uniform. I felt very downhearted but the Sergeant Major saved the day. He found a man who was not shooting at the time I would be on the range and it was arranged that we would change clothes. This we did and there must have been many smiles from the onlookers as we changed. I cannot remember my score or any more about that incident but it was good experience and rather funny.

A year or two after I joined the Volunteers, the Territorials came into being and the battalion was re-named the 9th Argyll and Sutherland Highlanders. There was no change in uniform as far as I can remember but the first winter we had a ball in the Gibson Hall and as I could not dance at the time I was a wallflower. I had taken a girl to the dance who was much older and who was a keen dancer and I think I might have tried one dance with her but that was all. I felt very conspicuous with my scarlet coat, brass buttons and white spats and decided that I would learn to dance before the next ball.

It was decided that the Battalion must have an annual camp and this was arranged to be held at Dalinlongart near Sandbank on the Holy Loch. A special steamer took us to Sandbank pier from Craigendoran and I had a splendid holiday there under canvas. The weather was perfect and I well remember the skin peeling off my brow, as the Glengarry cap gave little or no protection. The only thing I remember clearly was the route marches that took place. A small section of Battalion scouts had been formed with two members from each company and I volunteered for the job.

We set off on a route march one day and after marching for a mile or two up Glen Masson it was decided to have an exercise. The Scouts were to go forward and inspect the country in front and then return and report. I do not know what the Colonel thought we could report as no enemy had been arranged for us. However, we set off in pairs up the Glen and after a mile or two my partner and I decided that we would have a rest. It was a hot day and apart from the presence of clegs and horseflies we were enjoying ourselves. I am not sure what came over us but we decided to climb to the top of the hill to have a better view of the countryside. When we looked around we could see no sign of the Battalion anywhere and eventually we decided to cross the hills and see what lay on the other side.

We had no map or compass and by now we had lost our bearings so we toiled on to the top and from there we saw a small clachan with a road running up the valley. We were hot, tired and hungry and so we decided to make for the houses. Of course, we had all our equipment including rifles with us. We were soon in the valley and the first thing we saw was a shop selling lemonade and biscuits. Never did lemonade taste better or biscuits so good as these on that summer day so long ago. By now we had forgotten that we were Territorials on an exercise and it did not take us long to learn that we were now in Glen Lean on the road to Glendaruel. We had taken our tunics off as we were soaked with sweat. These army tunics were made to withstand winter cold and quite unsuited for exercises in the summertime. We found out that we had four or five miles to go to reach Dalinlongart and so in due course we set off down the road. We were certainly a pair of happy wanderers and about seven to eight pm we arrived at the camp. We learned that the Battalion had returned to camp hours before and in next to no time we were in front of the Scout Officer to explain our absence. I have no doubt we both looked very sheepish and the young Officer was obviously relieved that we were safe and well. But inwardly I was laughing and thinking about many things. I had thoroughly enjoyed my day's adventure and it was undoubtedly the highlight of my first camp with the Territorials

I have reached the end of my recollections of my youth in Garelochhead and in conclusion 1 have written a short poem on the Gare Loch.

Set in a frame of mountains high,
The Gare Loch sleeping lies below,
As breaking dawn lights up the sky,
Her placid face begins to show.

Awakening like a child from sleep,
Serene and still with eyes alight,
The sun reflects on waters deep,
And sparkles in the growing light.

And on the hills the heathers grow,
Brown and withered and forlorn,
But life is there, and soon will show,
The purple buds of bells re-born.

And scent the air for miles around,
The Honey Bees in search for store,
Will sip the nectar they have found,
To carry home and come for more.

Yachts on the Loch the waters skim,
Small boats are rowed about for fun,
Some wade or paddle, others swim,
And fishers hope, to land the biggest one.

The colours of the trees now change from green,
To warmer coloured tint and shade,
The moors in reddish purple gleam,
The Sportsmen count the grouse their guns have laid.

The sun sinks low behind the western walls,
Now burnished by the rays of setting sun,
The gloaming follows and the darkness falls,
The Gare Loch sleeps and rest to all has come.

www.ingramcontent.com/pod-product-compliance
Lightning Source LLC
Chambersburg PA
CBHW080444110426
42743CB00016B/3266